Human Factors for the Design, Operation, and Maintenance of Mining Equipment

Human Factors for the Design, Operation, and Maintenance of Mining Equipment

Tim John Horberry

Robin Burgess-Limerick

Lisa J. Steiner

CRC Press is an imprint of the
Taylor & Francis Group, an **informa** business

CRC Press
Taylor & Francis Group
6000 Broken Sound Parkway NW, Suite 300
Boca Raton, FL 33487-2742

International Standard Book Number: 978-1-4398-0231-1 (Hardback)

Library of Congress Cataloging-in-Publication Data

Horberry, Tim.
Human factors for the design, operation, and maintenance of mining equipment / authors, Tim John Horberry, Robin Burgess-Limerick, Lisa J. Steiner.
p. cm.
"A CRC title."
Includes bibliographical references and index.
ISBN 978-1-4398-0231-1 (hardcover : alk. paper)
1. Mine safety. 2. Human engineering. 3. Mines and mineral resources--Equipment and supplies. 4. Mining machinery--Design and construction. 5. Mining machinery--Maintenance and repair--Safety measures. I. Burgess-Limerick, Robin. II. Steiner, Lisa J. III. Title.

TN295.H74 2010
622.028'4--dc22 2010017184

Visit the Taylor & Francis Web site at
http://www.taylorandfrancis.com

and the CRC Press Web site at
http://www.crcpress.com

Contents

Foreword

I am extremely pleased to have been asked to write the foreword for *Human Factors for the Design, Operation, and Maintenance of Mining Equipment*. The book seems to be particularly well timed because there is a growing appreciation within the industry that proper consideration of human factors issues with respect to mining equipment is of critical importance. For example, a quick glance at the lost-time injuries, high potential incidents, and disabling injury statistics collected in Queensland, Australia, or elsewhere shows the importance of systematically considering the human element in mining equipment. We continue to have too many fatalities and serious accidents in this area, and we can no longer ignore this component of the problem. Also, no other recent publication is available to the industry that contains human factors knowledge regarding mining equipment in a single volume; in a busy industry such as mining, having easily accessible, user-friendly, practical, up-to-date, and comprehensive information is vital.

In Queensland, the Mines Inspectorate is taking the lead with respect to human factors and has just completed a twelve-month project which has resulted in a mining-focused, human factors–based tool which will be used as an aid in the investigation of accidents. As noted in this book, human factors are also being examined in conjunction with the use of proximity detection and collision avoidance on mining vehicles to minimise the interactions of vehicles with each other and pedestrians. The Queensland Mines Inspectorate will continue to propagate safety information of this type through all of the avenues available to us, and publications such as this play an important role in this process.

One particularly pleasing aspect of the book is that it can be used as a guidebook when searching for information about a specific human factors topic, or it can be read from cover to cover to gain a more complete understanding of human factors issues. Because it is reasonably succinct, careful reading of the entire contents is strongly recommended; such reading will help provide an important appreciation of what "human factors" are all about and their critical importance to safety, health, and efficiency in the minerals industry.

Stewart Bell

Commissioner for Mine Safety and Health
Executive Director
Safety and Health Division
Queensland Mines and Energy
Australia

Acknowledgements

The authors are grateful to colleagues at the University of Queensland, the National Institute for Occupational Safety and Health, and elsewhere for their assistance in many ways. We specifically acknowledge peers who ably co-wrote three chapters in this book: Tammy Eger from Laurentian University, Canada (Chapters 6 and 7), and Jennifer G. Tichon from the University of Queensland, Australia (Chapter 11). Also co-authoring in Chapter 6, we are grateful to Janet Torma-Krajewski from the Colorado School of Mines (Climate), and James Rider (Dust) and Robert Randolph (Noise), both from NIOSH Office of Mine Safety and Health Research. Photo credit to Dr Gary Dennis for Figures 4.1B, 4.1C, 5.1, 5.2, 8.2A, 8.3, and 8.4. The invaluable backing and sponsorship provided by Stewart Bell, Gerard Tiernan, Trudy Tilbury from the Department of Employment, Economic Development and Innovation (Queensland Mines and Energy), and Grant Cook from the Queensland Mining Industry Health and Safety Conference were particularly welcome. Robin Burgess-Limerick's contribution was supported in part by the Australian Coal Association Research Program (project C18012). The book draws on research projects undertaken by the authors funded by the Coal Services Health and Safety Trust, CSIRO, and ACARP projects C16013, C14016, C11058, C18025, and C17033. Finally, we offer our greatest thanks to our families for their unstinting and continuing support.

The Authors

Tim John Horberry, PhD MSc MIEHF BA (Hons), is an associate professor (human factors) at the Minerals Industry Safety and Health Centre, The University of Queensland, Australia, where he is director of research. Dr Horberry has conducted research in transport and mining human factors in Australia, the United Kingdom, and the European Union. Tim has published his research extensively—including a recent book, *Understanding Human Error in Mine Safety* (2009).

Robin Burgess-Limerick, BHMS (Hons) PhD CPE, is a certified professional member of the Human Factors and Ergonomics Society of Australia. Robin also holds a fractional appointment as associate professor in the School of Human Movement Studies, The University of Queensland, where he coordinates postgraduate programs in ergonomics. Robin has published more than fifty papers in international journals, and is the immediate past president of the Human Factors and Ergonomics Society of Australia.

Lisa J Steiner is a certified professional ergonomist in the United States. She is team leader for the Human Factors Branch at the Office of Mine Safety and Health Research, National Institute for Occupational Safety and Health, Pittsburgh, PA, United States. She specializes in safety and injury prevention, particularly with reference to mining environments and equipment along with facilities planning and design. Lisa has her master's degree in industrial engineering and is currently undertaking her PhD within the School of Human Movement Studies, The University of Queensland, Australia.

chapter one

What is human factors, and why is it important for mining equipment?

1.1 What is "human factors"?

Everyone has some idea of what is meant by "human factors." But to formalise things, and to provide some common ground, *human factors* (or *ergonomics*) is defined here as the scientific discipline concerned with the understanding of the interactions among people and the other elements of a work system, and as the profession that applies theory, principles, data, and methods to design in order to optimise human well-being, safety, and overall system performance. Or, as the vision statement of the Human Factors and Ergonomics Society of Australia (2010) puts it, "People-centred environments, products and systems for all."

As well as a scientific field and profession, human factors is also a way of looking at the world which has as its focus the capabilities, limitations, motivations, behaviours, and preferences of people. The aim is to maximise efficiency, effectiveness, quality, comfort, safety, and health by ensuring that systems are designed in such a way that the interactions are consistent with people's capabilities, limitations, motivations, behaviours, and preferences. The emphasis is on changing work systems to suit people, rather than requiring people to adapt to these systems. As the topic of this book is mining equipment, the emphasis is on designing this equipment and the broader work system to fit mining operators and maintainers. This can be achieved by measures such as through more fit-for-purpose initial design of equipment or by improving work procedures and processes.

The human factors philosophy implies a scientific concern with obtaining information about the characteristics and capabilities of people, and this is achieved by drawing on knowledge from the underlying disciplines and fields, including anatomy, physiology, biomechanics, anthropometry, neuroscience, social psychology, cognitive science, organisational psychology, management, work study, epidemiology, public health, and sociology. Another concern of human factors lies in design-based disciplines and fields such as product design, engineering, architecture, and computer science, where the opportunity exists to influence the design of systems, equipment, and environments. Firmly within that framework, this book addresses the role that human factors can play in the design, use, and maintenance of mining equipment.

1.2 *What are the aims of human factors?*

Human factors can therefore be thought of as having two separate (but not competing) aims:

- To improve work performance. This includes quantity of work (e.g., increased human machine integration leading to more ore moved), quality of performance (including accurate fault detection in maintenance work), and fewer errors, accidents, and/or near misses.
- To improve the health, safety, and well-being of the workforce and, where possible, the wider community, for example through fewer occupational injuries or health problems (e.g., noise-induced deafness), less job-related stress, increased usability of products, and increased work motivation.

Of course, the underlying assumption is that using human factors data, principles, and methods will lead to better designed jobs, tasks, products, or work systems. This is often easier said than done. However, this central theme (in various guises) will be present throughout this book.

1.2.1 *But ... people differ in shape, size, ability, skill, and motivation*

This is true whether that is in terms of mental processes such as decision-making or reaction time, or physical features such as the length of a person's reach. A key principle is that equipment, systems, and tasks should be designed to accommodate as many potential users as possible. One way to achieve this is through making parts of the equipment interface (e.g., a seat or even a display screen) adjustable; however, adjustability is only beneficial when it is done correctly to suit different individuals.

1.2.2 *And ... adding human factors to the design of a product is often seen as unnecessary*

This can be for a variety of reasons, including the following:

- Designers sometimes think they can use their knowledge, common sense, and intuition instead, or they rely purely on designing to standards.

- Adding human factors may be thought to alter an agreed-on design process.
- Similarly, an older version of a system may already be in place, and piecemeal alterations are subsequently made. For example, consider the road system around a mine site: it might be almost impossible to simply stop using it and design it again from scratch using human factors principles.
- The benefits and costs of using a user-centred approach for equipment design are not clear.

All these issues will be considered in more depth throughout this book. Needless to say, our view is that human factors is a key aspect to consider in the design and use of mining equipment.

1.3 *Why is it important to consider human factors for mining equipment?*

Based on the twin aims of human factors presented above, two broad reasons are investigated here. Both are critically important.

1.3.1 *Safety and health*

As we will explore throughout this book, human performance problems can constitute a significant threat to system safety in mining. It should be noted that in this sense mining is not unique, as other domains such as road transportation, defence, manufacturing, and maritime operations have broadly similar issues (Grech, Horberry, and Koester, 2008).

Except in the utopian dream of a fully automated mine, humans—whether operators, maintainers, trainers, supervisors, or managers—are a central part of the mining system rather than an optional extra (Simpson, Horberry, and Joy, 2009). Therefore, a full understanding of accidents or incidents will only be obtained if the human element is viewed as a fundamental part of a wider work system, rather than considering the system to be basically safe if it were not for the human part (Dekker, 2006). Despite this, looking at incident or near-miss statistics (especially incident narratives) indicates that human-related issues are somehow linked to a majority of occurrences (Patterson, 2008). In a complex system such as mining, it is usually far too simplistic to say that a single human failure "caused" the incident (as they usually have multiple causes, especially if probed far enough upstream), but they do at least indicate the importance of humans to the overall safety of mining work (Simpson et al., 2009). Another way to look at it would be to consider how often mining personnel "caused safety"—for example, by their actions they "caused" a task to be safely performed 999 times out of 1,000—rather than focusing on the single incident where things somehow went wrong. Further discussion of this issue is beyond the scope of this book; for the moment, the important points are that humans are central to promoting and maintaining safety, and that people are also involved in some measure in a large majority of mining incidents and accidents.

1.3.2 Productivity and work efficiency

It is relative easy for the human factors professional to play the safety card in mining, and to focus on human performance to improve safety (whether the standpoint is one of viewing human error as a cause of incidents or instead seeing it as a consequence of a problem in the wider work system). Whilst vital, the other aim of human factors, increasing work performance and productivity, is also a key area in which there is certainly plenty of work to be done in mining.

Partly this is due to organisational barriers, where it might be easier for a mining company to employ a human factors specialist in a safety department than in a job that generally helps to improve performance. Also, it is partly an artifact of what is being measured—the role that mining personnel play in safety is often easier to grasp compared to tackling how human factors might be used to improve productivity and efficiency.

Consequently, this area is often left to management consultants, business improvement consultants, organisational development specialists, or production staff. Whilst these can undoubtedly have a major impact, this situation underestimates the potential benefits of applying human factors principles to improve productivity. For example, as will be explored in more depth later in this book when considering new technologies, operators and maintainers are an essential component in overcoming many of the limitations of some mining equipment. They often adapt to make the technology work successfully, and whilst this can get the job done there are many instances where tasks could be done better, quicker, more efficiently, or more productively had human factors principles been more systematically considered. Other possible examples include a potential 5 percent gain in minerals industry process control efficiency through having better designed predictive displays (Thwaites, 2008), less wastage or errors due to a mine site having a systematic operator fatigue management program, redesigning jobs to require less manual tasks, or fuel savings in mobile equipment through having a fit-for-purpose operator interface. So a key mission for human factors specialists in mining, and for this book, is to sell the productivity benefits of using a human factors approach.

1.4 History of human factors in mining

The application of human factors to mining has a rich but uneven history around the world. The US Bureau of Mines, the British Coal Board, and the South African Council for Scientific and Industrial Research are all good examples of organisations that have conducted or funded human factors work related to mining. But things change: for example, in the UK the virtual collapse of the coal industry around the 1980s resulted in a serious decline in the amount of British work in mining human factors.

Much of the early published work is now difficult to track down directly; this is a particular problem because much of it was published over twenty years ago and therefore digital copies of reports or papers are not available. However, many important publications do survive, such as a US Bureau of Mines–funded book about human factors in mining by Sanders and Peay (1988) and the British Coal work of Simpson, Mason, and colleagues (partly summarised in Simpson et al., 2009). Further, the MIRMgate website of the Minerals Industry Safety and Health Centre at the University of Queensland (www.mirmgate.com; MIRMgate, 2010) hosts many of these older publications.

Recently, organisations such as the Australian Coal Association Research Program and the National Institute of Occupational Safety and Health (NIOSH, United States) (NIOSH absorbed two laboratories of the Bureau of Mines when the Bureau of Mines was eliminated) have actively funded research in the area, often within a broader health and safety framework. Their role in providing a continuity of research funding has helped allow experienced human factors researchers to remain in the field. Their open dissemination of the research outputs is especially welcome: it is a benefit to all of the industry and an additional source of key material for this book.

1.5 Human factors and risk management

As mentioned above, a risk management framework is adopted in contemporary mining human factors to guide the application of the principles and knowledge to any particular equipment design problem. The process starts with establishing an understanding of the broader context in which the particular person–equipment interaction takes place before undertaking hazard identification and risk assessment. Assuming that the outcome of the risk assessment is that action is indicated, the risk control phase incorporates identifying and evaluating potential control options, before implementation and ongoing review. From the human factors perspective, as will be seen in Chapter 2 of this book, the emphasis for risk control is on elimination or reduction of risk through design controls rather than focusing excessively on administrative controls such as training, selection, or personal protective equipment.

Most importantly, this process also places emphasis on consultation with the people concerned at each step. This issue is at the heart of "participative ergonomics" approaches, which take as an underlying assumption the notion that the people involved are the "experts" and must be involved at each stage of the risk management cycle if the process is to be executed successfully. In an occupational injury risk management context, this implies in particular that employees and management participate through hazard identification, risk assessment, risk control, and review steps of the risk management cycle. Evidence exists to demonstrate the effectiveness of such approaches across industry in general and in mining in particular (Burgess-Limerick et al., 2008; Torma-Krajewski et al., 2009). Such a participative approach can also be used for productivity and work efficiency gains by improving equipment design, procedures, and training.

1.6 *Key current issues, and future challenges with mining equipment*

The above has defined human factors and made a general case for its greater deployment in mining. To build on those points, this section briefly outlines five current or emerging human-related issues for mining equipment.

1.6.1 *Safety versus production*

This first area is, of course, not just restricted to mining equipment, as it is an issue across all areas of mining. Indeed, the balancing of safety against productivity is an issue for most domains. But to focus purely on mining equipment, different trade-offs certainly do exist: for example, load size versus machine stability, or equipment speed versus incident likelihood and/or severity.

Trade-offs exist at all stages of the mining equipment life cycle, including at the equipment design stage, during equipment procurement and purchasing, during maintenance (especially scheduled maintenance), and during equipment operation. Training of new operators with equipment is not immune to trade-offs, either: how much and what type of training is given? More training might have safety benefits (plus longer term productivity gains), but this needs to be offset against the costs of training, the possible need to take equipment out of service whilst training takes place, and the need for experienced operators to actually give the training. Of course, offline training such as using mobile equipment simulators can reduce some of these effects, but they are certainly not without costs to a mine site.

Sadly, of course, no magic solutions are offered here to resolve this safety versus productivity trade-off issue. But greater efforts to identify, recognise, and assess where the trade-offs exist, and using that as a basis for informed decision making, comprise the main recommendation here. In this book, many examples of such trade-offs with respect to mining equipment will be shown, for example regarding improved visibility versus reduced functionality of some underground equipment.

1.6.2 Bigger! Stronger! Quicker! Safer! More reliable!

There is an increasing demand for mining equipment parameters to be improved. Compared to a few years ago, there is an expectation that mining equipment will be bigger, stronger, quicker, safer, and more reliable. For example, the current size of shields in underground longwall coal production would be unimagined a few years ago. Similar issues exist in surface mining with mobile equipment (see Figures 1.1 and 1.2 for surface mining, and Figure 1.3 for underground mining).

Size is not the only important parameter. High equipment reliability, more robust designs, and efficient performance are the expectations, or at least the anticipations, of mines and mining companies. This, of course, imposes pressure on equipment manufacturers to design and build to

Figure 1.1 Example of the size of equipment used in modern mining.

Figure 1.2 Another example of the size of equipment used in modern mining: truck with five coal trailers.

Figure 1.3 Longwall shield.

such requirements, for site-based maintainers to service such equipment safely and efficiently, and for operators to be actually able to operate it optimally. However, with increased competition and economic pressures, it seems unlikely that such demands will reduce in the future.

1.6.3 Remote control and automation

It also seems likely that there will be more remote control and/or automation of mining equipment in the future. This may change the types of human factors inputs required (for example, less manual operational tasks—at least when the equipment is working correctly), but it certainly does not remove the need for human factors involvement.

As will be seen in Chapter 9 about new technologies, some of the lessons from other industries where remote control and automation have already been deployed on a larger scale (e.g., defence or aviation) show that operator jobs certainly do change (often to more of a passive monitor of the system, rather than an active controller or driver of it). Indeed, this can create potential problems in that if the "passive monitor" operator of an automated system loses situation awareness, and/or becomes deskilled, he or she may be unable to take appropriate corrective action in the event of equipment malfunction. Experience in other industries also demonstrates that human factors issues such as how the equipment status information is displayed (both for operators and subsequently for maintainers), how it will be controlled, and how acceptable it is to personnel are of key importance, as well as what happens if the system malfunctions. Neglecting these issues will often result in equipment safety and performance problems, such as improper use, employee sabotage, or at least employee distrust.

1.6.4 An ageing workforce

Twenty years ago, it was noted that the average age of the mining workforce was getting younger (Sanders and Peay, 1988). This situation has reversed today, and the mining workforce in most industrialised countries is on average now getting older, and often more overweight. Although this mirrors many general trends in society, it does present a few specific issues for the design, maintenance, and operation of mining equipment. Although extreme ageing is not an issue in mining, there are still many issues for an older workforce. These are revealed by highlighting the general effects of ageing:

- There often is an increased difficulty in learning new skills. Older people do not automatise tasks so easily compared to their younger counterparts. This has implications for mining technology use where skill requirements may change over time and require new automatic, over-learnt operating procedures.

- Most memory functions decrease in performance.
- Reaction time increases, especially an increasing delay in reacting to unexpected stimuli.
- Loss of muscular strength, endurance, and tone, especially in the lower extremities.
- Bones weaken and become more porous and less dense.
- Stiffness and joint pain (e.g., arthritis) become more prevalent.
- Respiration capabilities are reduced—lungs are less efficient and more prone to disease.
- Blood supply (especially to extremities) is diminished. The size of the heart muscle may be reduced, and the heart rate takes longer to return to resting level after exertion. This is exacerbated if significant weight gain occurs in older workers.
- Visual function changes include loss of precision, and difficulty in focussing on near objects. For example, there is a link between ageing and light wavelength, this can be a problem with LEDs geared toward blue-green wavelengths as older workers might not see them as well as more neutral wavelength LEDs.
- Visual acuity and contrast sensitivity decline, especially in bad lighting conditions or with a complex background.
- Similarly, hearing decreases in most people, first for high-frequency sounds and then for lower ones.
- Pain, touch, temperature, and vibration are more difficult to detect. Likewise, the senses of taste and smell often diminish.

However, it should be noted from this depressing list that many changes are actually more related to fitness than to age. For instance, it is not uncommon for a fit fifty year old to be stronger, be slimmer, and have the same reaction times as a sedentary thirty year old. Similarly, older workers often have compensation strategies (such as tactics to memorise things or not performing certain activities). They may also have better perceptual expertise, and be less likely to take unnecessary risks. However, the above-mentioned physical, perceptual, and cognitive declines with increasing age can still present a significant challenge to equipment designers, workers, and mine managers.

1.6.5 Gap between mine site ergonomics knowledge and manufacturer human factors design skills

The final issue to be raised here is about gaps and disconnects. Larger mine sites often have personnel with reasonably good general knowledge of physical aspects of human factors and ergonomics; however, generally they have less specialist skills in human factors, especially in the more "cognitive" areas. Conversely, equipment manufacturers, of course, need to have personnel with good design skills. Often these personnel can very competently design to existing human factors standards (at least, where they exist). However, they sometimes do not have extensive site experience, and may not have significant "voice of the customer" knowledge or participatory ergonomics skills. This gap was one of the motivations for the formation of the Earth Moving Equipment Safety Round Table, a recent international initiative that is further described in subsequent chapters in this book.

Added to this is the disconnect that often exists between what happens (and what is known) at a mine site level and what is known at a corporate level for some mining companies. This issue is wider than just related to human factors issues of mining equipment, and it can have impacts for mining equipment in terms of equipment procurement, training policies, and the development of safety procedures and protocols.

1.7 Why this book is necessary

There is increasing international recognition within the mining industry about the importance of considering human factors issues in the design, maintenance, and use of equipment. This is evidenced on several fronts:

- During incident investigation, site personnel becoming increasingly aware of the strengths and weaknesses of their operators and maintainers. This is accompanied by examining the behaviour of these operators and maintainers within a wider work context (including equipment design and use), and so not automatically subscribing the cause to "human error" that simply requires "more training."
- Equipment manufacturers beginning to give systematic attention about designing to human abilities (and often making their attention to ergonomics a particular selling point).
- Funding bodies increasingly supporting human factors and ergonomics research programs.
- Mining company managers at a corporate level regularly talking about better harnessing the human element in future mining systems
- Regulators paying particular attention to systems in place to manage person-related safety.

This book is timely because there is no recent publication available that focuses on a wide range of human factors knowledge and research related to mining equipment,* and it provides a much-needed overview

* However, recent books are available about either mining human factors (e.g., Simpson et al., 2009, which focuses on human error) or mining equipment from more of an engineering and management perspective (e.g., Dhillon, 2008).

of the human element. It will be useful for many professionals in the field; these include mining companies, regulators, health and safety personnel, equipment manufacturers, designers, engineers, researchers, and students. Indeed, the book has been written to appeal to all these groups by providing a balance between breadth and depth in the treatment of different areas, and pointers and additional references are given to help the reader explore in more depth particular topics of interest.

1.8 Structure of the book

To help set the scene for later chapters, the book begins by presenting fairly general, yet fundamental, human factors information related to equipment. In this vein, Chapter 2 focuses on mining equipment design; this introduces key information in this area such as the notion of the equipment life cycle and the "hierarchy of control." A case study about mobile equipment is outlined at the end of the chapter; this helps to bring together many of the concepts introduced. Chapter 3 presents the other side of the situation, and shows why design is not the only issue: mine site rules, procedures, work methods, guidelines, and similar topics also have significant impacts upon mining equipment safety and efficiency. As with Chapter 2, several key concepts such as "Haddon's matrix" are introduced and illustrated in this chapter.

Chapter 4 is the beginning of the more specific human factors content. The material is grouped so that Chapters 4–7 cover more physical issues related to mining equipment. In more detail, Chapter 4 focuses on manual tasks currently performed by operators and maintainers and the problems that may result. Chapter 5 looks at sizes and spaces: sizes of people, workstation design, access and egress problems, and issues with working in confined spaces.

Next, Chapter 6 considers the physical environment, and how factors such as vibration, noise, heat, and dust are of key concerns for people maintaining and operating mining equipment. Then, Chapter 7 examines vision, lighting, and equipment visibility, and ways to enhance visibility (or at least cope with a lack of visibility).

The focus then switches: Chapters 8 and 9 examine more "psychological" aspects concerning mining equipment, including how information is perceived and processed, and how people's actions are planned and undertaken. Chapter 8 concerns equipment controls and displays, the so-called human–machine interface, and it outlines many of the human factors issues of importance in this area. Chapter 9 builds on this by examining new technologies and the greater use of automated mining equipment. Drawing on lessons learnt in other domains (e.g., aviation), it shows some of the benefits and pitfalls of advanced technologies, and why a systematic consideration of the human is still very much required.

Chapter 10 then expands the scope by examining wider organisational and task factors related to mining equipment. Amongst the issues considered are long-standing problems of operator fatigue and stress, as well as newer concerns such as distraction and information overload. Chapter 11 continues the broader organisational theme by focusing on training. It gives an outline of the area from the first principles of skill acquisition to how simulation can be used in mining equipment training.

Chapter 12 ends the book. It brings together much of the previously introduced material and makes the point that the consideration of human factors within mining equipment is certainly not a static field. The concept of maturity models is used to show how both equipment designers and mine sites are on a journey towards more optimal equipment design processes and end products.

chapter two

Equipment design

The design of mining equipment plays a crucial part in the safety and efficiency of work tasks that are conducted by operators at that equipment. Similarly, design has a major impact upon the ease, safety, and efficiency of equipment maintenance. This chapter focuses on equipment design, looking at key aspects such as the design process, the equipment life cycle, and human factors issues concerning safety and usability. Given that the target audience for this book includes equipment designers and manufacturers, mining site personnel, safety professionals, as well as students, researchers, and others, the information contained here needs to be reasonably general. At the same time, it sets the scene for Chapters 4–11, which deal with more specific human factors issues related to mining equipment.

2.1 The equipment design process

On the broadest level, one of the main things all equipment designers have in common is a design process, that is, the activities they go through whilst conceptualising, designing, and building a new piece of equipment. Of course, the actual process differs according to different manufacturers, individual designers, the type of equipment being designed, and external constraints such as time and budget.

There are many ways of picturing the design process, and usually these are a generalisation of what really happens. At the simplest level, the process ranges from concept, through to pre-feasibility of the design, then feasibility, then a first-prototype build, then an advanced-prototype build, and then manufacturing and commission.

Perhaps the most important thing to note about the design process is that it is iterative—it is not a fully linear process where a single idea remains essentially unchanged from concept through to production. For example, upon building a first prototype the designer might return to the original concept and modify this. So the process, especially for more complex mining equipment, usually involves revisiting, refining, and changing design ideas, and testing, evaluating, and refining are virtually always necessary.

Human factors should play a major part in this whole process. This includes major criteria in the testing process (e.g., usability testing, as will be reviewed later in this chapter) and also in providing input into different stages of the design process (e.g., user acceptance). It is not within the scope of this book to go into excessive detail about how human factors information and methods can be used in overall design, and other textbooks cover this area well already (e.g., Chapanis, 1996). However, as will be seen throughout this book, key human factors information includes how user requirements are captured, how human factors information is provided in an accessible form for designers to use (e.g., regarding human sizes and strengths), how appropriate human-centred methods are used to evaluate designs (e.g., user trials), how training needs are analysed and appropriate training provided, and how the equipment is integrated into the work system (e.g., looking at possible resistance to change issues, how procedures need to be modified, or where equipment retrofits are needed).

However, one of the "paradoxes" is that although human factors and ergonomics should certainly be involved early and often in the design process, the certainty of the effects of design changes on safety and performance of the actual equipment is often not fully revealed until the equipment is operational and the exact context of the working environment and work tasks are known (Hendrick, 2003). Of course, when the equipment is operational it is "too late" for the design to be changed (at least for that exact equipment model), hence the design of much mining equipment often requires small modifications and improvements from one exact model to the next one.

2.2 *The equipment life cycle*

The design process section above describes some of the stages that a designer would go through, from initial concept through to manufacturing of the final equipment, and hopefully extending to also consider operation, maintenance, modification, and disposal of the equipment. Very much linked to this is the notion of equipment life cycle. Again, there are many variations on the basic theme, and it varies according to equipment type, industry, and external constraints, but for most mining equipment it can be shown as in Figure 2.1.

As was seen in Chapter 1, some of the main drivers in mining equipment are the requirements to design, build, buy, run, and maintain equipment and technologies that increase the capacity for productivity as well as improve the safety and health of the workforce. However, human factors knowledge and methods are not currently considered as part of the core issues in mining equipment design, operation, or maintenance. For example, looking at the earlier stages in the system life cycle, since design engineers develop most human–machine systems, and human factors is not normally in their toolbox of design methods, there is a tendency to focus more on technical aspects of what the mining equipment can do. Often they assume that the tasks the equipment cannot do

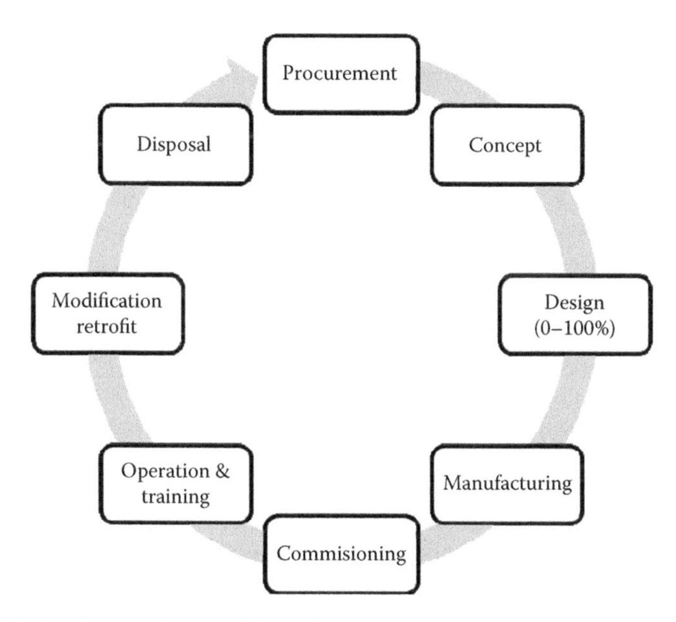

Figure 2.1 Mining equipment life cycle.

would be fulfilled by the human operators and maintainers. As will be seen in Chapter 9, this is not an appropriate assumption.

Because most equipment designer and mine site engineers still view human constraints to be less significant than technical challenges (such as of equipment reliability, payload, etc.) there is a tendency to not systematically consider human factors in the equipment life cycle process, and it is common to see human factors concerns being passed from one phase to the next. For example, if during conceptual design the requirements analysis does not adequately capture user requirements, then subsequent inadequacies cannot be resolved in the design phase. Problems that remain after the system has been designed cannot, therefore, be resolved during equipment build or implementation; this is a particular problem for mining equipment that requires considerable human intervention in its maintenance and operation (e.g., for continuous miners underground).

Retrofitting and modification might occur, whereby human factors specialists may be brought in to solve specific, and typically serious, equipment acceptability or usability problems once the equipment has been implemented. This, however, is usually far from ideal since such changes are often more difficult, less effective, and much more expensive at this point, and may be impossible. Worse still, equipment issues that are not dealt with by a human factors specialist may be addressed only by the system documentation, or training of operators, maintenance, and support personnel. Most people reading this book will have already experienced situations where they have required training or lengthy documentation to cope with the inadequacies of mining equipment design.

On the flip side, and to share the blame for this situation, human factors professionals are often more traditionally interested in analysis than in design, and have a different culture and professional language (e.g., workload and situation awareness) than designers or mine engineers. Communication can be ineffective; one of the functions of this book is to try to help bridge that gap by providing human factors information that is applicable to mining equipment design, operation, and maintenance in one accessible source.

The third party in this process is the customers. Here the blame is shared also. If mining companies place unrealistic demands, especially time demands, on manufacturers, the probability that the designer will adequately consider human factors principles within each design phase is reduced. On the positive side, at least one multinational company is now requiring tenders submitted by manufacturers to describe how human factors principles will be incorporated within the design process, and in particular at what stages and how user feedback will be incorporated in the design. This tender process also requires a broad brush risk assessment related to human factors issues to be conducted by short-listed manufacturers prior to contract award, and for a more detailed risk assessment to be conducted by the site with the assistance of the manufacturer prior to release of the equipment drawings for manufacture. In theory, this process should ensure that the major human factors–related hazards are identified, and controls determined, much earlier in the design phase than has previously been the case.

2.3 *Safety in design*

Incorporating human factors early and often in the equipment design process and life cycle is therefore one of the central arguments in this book. One way to broadly achieve this is through *safety in design* (sometimes known as *safe design* or *prevention through design*). This approach has received a great deal of recent attention in both the scientific and occupational safety domains (including in mining) and is generally applied to products and equipment. As the name suggests, it involves safety by design, not safety by procedure or through retrofit trial and error. A more complete definition of the topic, according to Safe Work Australia, is as follows:

> Safe Design is a design process that eliminates OHS hazards, or minimises potential OHS risk, by involving decision makers and considering the life cycle of the designed-product.

> A Safe Design approach will generate a design option that eliminates OHS hazards and minimises the risks to those who make the product, and to those who use it. (Safe Work Australia, 2009)

In equipment design (especially mobile equipment design), a distinction is often made between primary, secondary, and tertiary safety:

- Primary safety is the prevention of accidents per se.
- Secondary safety is the protection of the person in the accident situation: for example, making mining vehicles more crashworthy.
- Tertiary safety involves recovery and assistance after an accident.

Safe design should also therefore consider all three levels; however, the main concern of eliminating hazards and minimising risks means that the focus is on primary safety, especially for the users of the equipment.

The case study that is presented at the end of this chapter concerning the development and use of the operability and maintainability analysis technique (OMAT) will illustrate the safe design concept in a more concrete manner, and will show the links between safety in design and human factors. OMAT also relies on the notion of hierarchy of control, and this is the topic of the section below.

2.4 Hierarchy of control, and control effectiveness

To improve safety, mining equipment–related hazards need to be controlled and managed. Some strategies previously applied to the control of such mining equipment hazards include elimination or reduction of the hazard, removing people from the hazardous environment, isolation of the hazard, use of engineering controls, application of administrative controls, erecting warning signs, use of personal protective equipment, and application of behavioural methods (e.g., disciplining operators). One categorisation method for all these is known as the *hierarchy of control.* Many variants of this taxonomy exist, and different mining companies and equipment manufacturers use different variations of the hierarchy. Perhaps the simplest version has just three levels; in order of likely effectiveness, they are as follows:

1. *Remove the problem*—by designing it out.
2. *Place a barrier around the object*—to stop the problem from occurring by placing a barrier (physical, organisational, or temporal) around the hazard. This might include a physical guard around the equipment to protect against accidental contact, or an organisational

Figure 2.2 Warning from a continuous miner.

guard where only qualified personnel work near that equipment (e.g., maintenance workers).

3. *Implement warnings*—to provide information needed to function safely with the equipment, or provide training; for example, placing notices near the equipment (see Figure 2.2) to indicate a particular hazard (e.g., pinch points).

In terms of the likely effectiveness, in this version of the hierarchy, warnings are placed at the bottom of the safety hierarchy, behind design and guards. The reason for this is that people may not see or hear warnings, they may fail to understand the warnings (especially if their reading abilities are low), or they may not be motivated enough to pay attention or comply with warnings.

Recently, more detailed and sophisticated versions of the hierarchy have been developed. For example, a seven-step version shown in Figure 2.3 has been used by Horberry et al. (2009) in which design to eliminate the energy source is the top of the hierarchy.

As an example regarding the second level, substitution and minimisation, measures could be put in place such as using an electric rather than an engine-powered freight mover (such as a forklift truck), as shown in Figure 2.4.

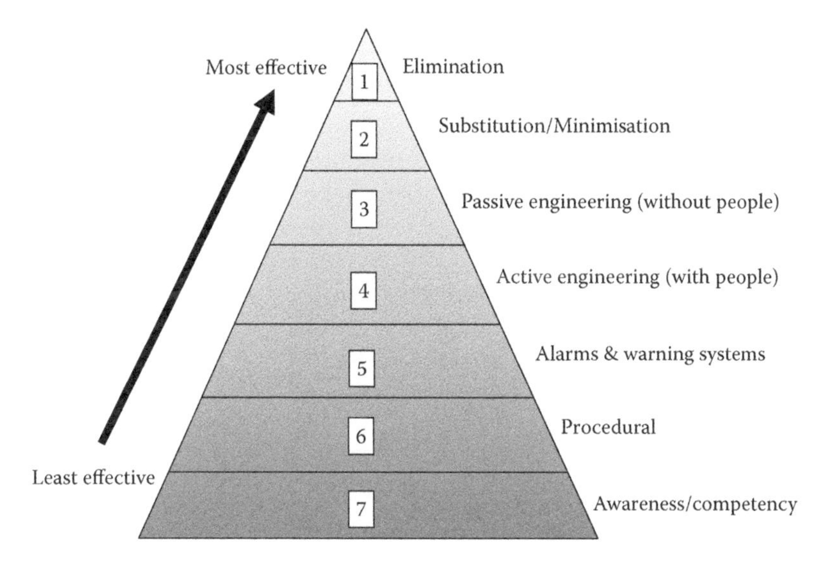

Figure 2.3 A seven-step level version of the hierarchy of control (Adapted from Horberry et al., 2009).

Figure 2.4 Substitution: Using an electric rather than engine-powered freight mover.

2.4.1 *Is the actual effectiveness of the controls more important than where they sit on the hierarchy?*

One issue that has been discussed recently is the effectiveness of the controls. The argument states that it is better to have more effective lower level controls rather than less effective higher level ones. For example, if a hazard can be controlled more effectively by competent people and successful procedures (i.e., the bottom two levels of the seven-level hierarchy) than by the midlevel control of active engineering which is preferable? This is currently an area of considerable research interest in the safety and risk management community, but a correlation between level in the hierarchy and effectiveness of the control seems certain to exist.

2.5 *Equipment usability*

Switching the focus back to more traditional human factors concerns, usability was mentioned earlier in this chapter as a key issue in mining equipment design. To provide a simple definition, *usability* means that the people who are intended to use a specific item of mining equipment can do so quickly and easily to accomplish their required tasks. Everything to do with the equipment has to be usable (Grech et al., 2008). This refers to not just sophisticated new mining technologies but also their documentation and maintenance. Similarly, usability applies equally to simple mining equipment such as hand tools.

In many ways the benefits of usability are obvious; they can include increased mining efficiency, fewer work injuries, improved operator acceptance, fewer errors, earlier detection of the requirement for maintenance, less training required, and less mistakes and/or violations. Mining technology should not be seen as a "barrier," whereby the operator has to focus excessively on the technology and not the task. Likewise, as much as possible, the technology should fit the operator and the task, and not require the operator to fit the technology. A recent example of this is the development work taking place in the area of remote control of mining equipment by many equipment manufacturers, mining companies, and scientific researchers. With newly introduced remote control systems, operators often need to adapt and excessively focus on the remote control interface rather than the actual task at hand. It might be argued that non–line of sight remote control of equipment, such as a continuous miner in an underground coal mine, is such a radical and promising change that a certain amount of learning and adaptation is unavoidable. Whilst this might be so, developing better and more intuitive remote control interfaces in the first place is still preferable. This whole issue will be explored further in Chapter 9 in the discussion about mining technologies and automation.

The application of human-centred design principles can help to ensure that the final product is usable. At present in mining, more formal usability testing of equipment is often undertaken only by the manufacturer and developer rather than the purchasing organisation or an independent human factors practitioner. Although manufacturer-based usability testing is necessary, it is not always sufficient—as specific mine site issues can have a significant impact (e.g., assessing the needs for new equipment in the actual site work environment, where climate, site procedures, and what other equipment is also employed can greatly affect equipment usability). It is also possible that user expectations for the way in which equipment functions or is controlled might be influenced by experience with equipment already on site.

Despite this, whoever undertakes equipment usability testing should consider the following:

- Identifying who are the operators or maintainer, and their likely capabilities and limitations—physical (e.g., their size), perceptual/motor, and cognitive (e.g., their reading skill levels)
- What the equipment should be used for
- How the operator will actually use it, and what might be foreseeable misuse (such as standing on a piece of equipment to undertake a brief working-at-height task, rather than using an elevated work platform)
- What other systems, equipment, and processes are in place and how the equipment will integrate with these existing items
- How its successful design and deployment will be evaluated, and how any subsequent lessons will be learnt

2.5.1 *Who benefits from a user-centred focus?*

The return on investments from using human factors is often long term and should ideally only be evaluated over the full life cycle of the equipment, rather than as "quick-and-dirty" measures to provide corrective and/or retrofit action and an apparent rapid return. When promoting the idea of employing human factors into the equipment's life cycle, many questions can arise:

- How much will mine site personnel, or the mining organisation itself, benefit?
- How will we know what are the benefits anyway? Will it be better safety, improved productivity, fewer errors, reduced training time, and/or greater equipment acceptability to the workforce that offer the main benefits?
- How much will it cost to integrate human factors into the design process?

- How much will be gained in terms of revenue or safety?
- How can we persuade company senior management to do it?

Such questions lead nicely into the discussion of human factors cost–benefit analysis and the system life cycle. Traditionally this has been an area where human factors professionals have only dabbled, and further work here is vital to help "sell" the worth of applying human factors to mining equipment manufacturers and mining companies.

2.6 *Human factors cost–benefit analysis and the system life cycle*

To recap, it has been argued above that human factors information and methods are usually best applied early and regularly throughout the design and development of mining equipment. However, many important human factors interventions can also take place at other times, such as post-implementation retrofits or modifications.

Whatever the time in the system life cycle, human factors interventions in most industries (including mining) usually have the dual objectives of increasing productivity and effectiveness and improving the conditions of work (for operator safety, health, comfort, and convenience). Although consideration may be given to ease of use, comfort, and other similar matters, most of the analysis has been done on both short- to medium-term performance and safety benefits. Such an approach does not usually fully consider long-term environmental costs (e.g., environmental pollution due to human error when maintaining equipment). Therefore, the complete benefits from using human factors should ideally be derived from the total system and equipment life cycle costs of the product, including conception, design, development, build, purchasing, implementation, support, operation, maintenance, and disposal.

By systematically outlining the benefits and costs, cost–benefit analysis (CBA) may help the designers and manufacturers of the system, the purchasers of the system, and the actual operators and maintainers of the system to see overall positive value from the application of human factors engineering. However, the process is not always too straightforward, and as will be seen below, significant obstacles can exist.

2.6.1 *Problems with human factors CBA in mining*

In an ideal world, all the costs and benefits (e.g., performance and safety aspects) would be gathered together both before and at several points after employing human factors principles in the equipment design process, and a comparison could be made to assess how successful the intervention was. If an identical piece of equipment existed that did not have any human factors input, then this could be of additional use as a baseline. Also, if multiple cases of human factors CBA for different types of equipment existed, then this would further strengthen things. However, examining the topic in more detail shows that undertaking human factors cost–benefit analysis is often problematic. Amongst these difficulties are the following (adapted from Rouse and Boff, 1997; Grech et al., 2008):

- Benefits are rarely purely monetary (e.g., ease of use of a new equipment type, or improved operator working conditions). Ease of use can indirectly be translated into measures such as fewer errors, reduced training, improved equipment acceptability, and the like, but these are still often difficult to assign a monetary value to.
- Investments or costs of human factors interventions often occur long before returns are realised. Thus benefits are delayed, obscured by time, or no longer relevant. A classic example here concerns longer term health issues from mining equipment (e.g., hearing loss, or back injury) where the effects might take years to fully develop and by which point the individual concerned might have changed jobs or retired.
- Costs and benefits are often distributed amongst a wide range of stakeholders (e.g., for a new type of mining equipment such as automation of some form, this may include the operators, supervisors, maintainers, managers, company accountants, and even regulators). Thus input from all of these groups is often necessary.
- Benefits might be related to events that do not happen—such as mobile equipment accidents being avoided due to good human factors design in the operator interface. Aggregate data (e.g., accident statistics over a number of years for different equipment types) could be used, but these very rarely would have the level of precision needed to make firm conclusions.
- In most mining systems, the effect of improvements to one small part of the system (e.g., better controls and displays in one piece of equipment) might be methodologically difficult to assess. Similarly, side effects are likely: these might be in the same direction (e.g., fewer incidents and decreased training costs) but not always so.

2.6.2 Benefits of using a structured CBA method

Despite the sorts of difficulties mentioned above, Rouse and Boff (1997) argue that a structured method to assess cost and benefits of human factors can help. They report that where structured CBA of human factors initiative has been undertaken in other industries (e.g., military), the results show the significant economic advantages of employing human factors. For this, they argue that it is vital to define the types of human factors benefits as exactly as possible. They state that these benefits can range from the very tangible to the less tangible, and may include the following:

- Solution not otherwise possible (e.g., a piece of mobile equipment not being able to be operated)
- Acceptable performance or cost not otherwise possible (such as a vehicle being able to travel at higher speeds)
- Improved operator performance, efficiency, and/or cost savings (e.g., reduced power use)
- Enhanced customer willingness to pay
- Cost and mishaps avoidance (e.g., fewer errors during routine maintenance of the equipment)
- Better acceptance (operators are more likely to use and trust a user-friendly device)
- Increased confidence by all those involved in the system

So, to conclude, there are substantial challenges associated with cost–benefit assessment of human factors work; this applies as much to mining equipment as it does to commercial products or military operations. Applying a structured methodology that emphasises what should be represented in the analysis can improve the situation, and where this has been done in other domains using human factors the results are encouraging. Despite this, obtaining fully quantitative figures is either impossible or at best imprecise, but still should be attempted. Such figures can also be useful in comparing between alternative equipment design solutions.

2.7 *Equipment standardisation*

An approach to design that is often necessary but rarely sufficient for complex mining equipment is to standardise. This might include the location of underground equipment access and egress (and the location of emergency access and egress if one exists), which side of the road mobile equipment should be driven on (which has follow-on effects for the position of the steering controls), or the means of isolating the equipment. Having standardised designs means that future versions of the equipment would largely follow the same parameters.

One method of obtaining equipment standardisation is through the application of standards, regulations, and guidelines. Many standards and guidelines exist in most domains; for example, in the surface-mining domain the International Standards Organisation group for Earth Moving Equipment is ISO Technical Committee 127. In Australia, the New South Wales Department of Primary Industries (now renamed as Industry & Investment NSW) also publishes Mining Design Guidelines (MDG) to assist mining companies and equipment manufacturers in meeting their obligations to provide fit-for-purpose equipment.

2.7.1 *Issues with standards*

Most standards applicable to the minerals industry only deal with fairly simple environmental, equipment, and task stressors (such as equipment noise levels over a working day) rather than complex control activities involving new technologies. The number of standards and regulations that specifically exist for automation is low.

It is not within the scope of this book to review all possible standards concerning mining equipment, but often operators and maintainers are faced with equipment manufactured by several different companies—where each uses its own internal standards, causing a total system to lack consistency. For example, as will be considered in more depth in Chapter 8, controls and displays can vary greatly between mobile equipment designed to do the same job (e.g., between different types of haul trucks, or between different makes of roof botting equipment), so operators may need to adjust regularly to different layouts, and such variation is a fertile source of errors. In theory, having international (e.g., ISO) rather than purely national standards (e.g., American, Australian, or Chinese) allows multinational manufacturers to make one product that will be sold all over the world. Given the international aspects of most mining operations and companies today, this can produce large cost savings, and often has usability benefits for the end user.

In most mining operations, things are often more complicated: as well as standards and regulations, other good practices or guidelines with respect to equipment exist. Simple examples are maintenance schedules, equipment risk assessment procedures, training processes, and emergency procedures. These can radically alter how the equipment is procured, operated, maintained, and modified, so risk-based rather than prescriptive processes are often preferable (Horberry et al., 2009).

2.7.2 *The standards process*

Most standards bodies rely on unpaid assistance by the various committee members, and traditionally the representation on the committees comes more from equipment manufacturers or regulators rather than users. The ISO standards process is formal and not always too rapid, thus it can take a long time for new standards to appear (over five years is not uncommon, from our experiences on ISO committees). The NSW (Australia) MDG process can move a bit faster, and typically has better user representation. In fast-moving areas like collision detection for mining equipment, standards are often well behind the actual technology. In some cases this may result in rather ineffective or weak standards as they need to be written in general terms to accommodate possible future technologies. This general nature of some standards may not be too helpful for designers. Similarly, they may only establish minimum requirements (sometimes due to a compromise by the various committee members).

Agencies can also be lobbied out of introducing potentially beneficial prescriptions. For example, as noted in Chapter 8, the need to standardise controls on underground bolting equipment has been highlighted numerous times since at least 1973; however, a (US) Society for Automotive Engineers standard titled "Human Factors Design Guidelines for Mobile Underground Mining Equipment," which addressed these issues, amongst others, was defeated at a ballot in 1984. The standard was not issued despite meetings continuing until 1990. The issues were again canvassed in 1994 by a committee formed by the Mine Safety and Health Administration (MSHA), following three fatalities involving roof-bolting machines in a six-week period. MSHA subsequently called for industry comment on an advance notice of proposed rulemaking titled *Safety Standards for the Use of Roof-Bolting Machines in Underground Mines* (MSHA, 1997), which suggested that MSHA was developing design criteria for underground bolting machines; however, no related rule or design criteria were subsequently released. It was not until the release of a draft Mining Design Guideline 35.1 by NSW in 2006 that many of the issues identified many years earlier in the United States have been incorporated into a published guideline (Industry & Investment NSW, 2010).

All this presents a slightly bleak picture. To emphasise the positive aspects, it should be stressed that standards often can have safety and efficiency benefits by promoting design consistency. Likewise, standardisation can also fit well into the process of conducting audits for purposes such as to assess cost–benefits of different equipment retrofit interventions, can assist in equipment procurement, and can be helpful to focus future training requirements. Finally, the application of standards can also assist in mining regulation, compliance, certification, and insurance, and when developing and applying safety management systems and other policies. However, as this book will demonstrate, not every human-related aspect of mining equipment design, operation, and maintenance can be standardised.

2.8 Potential barriers to using human factors in design

So if human factors is such a great discipline and approach, then why is it not systematically used in all mining equipment design? The material in this chapter has already indirectly mentioned a few of the reasons; the points below will summarise some and expand on others.

1. Lack of human factors knowledge and over-imputation. That is, designers get some of their information about the eventual mine site user by attributing their own personal knowledge to the other person, rather than using formal human factors data or methods.
2. Lack of easily accessible human factors information and methods, and standardisation does not always work for all human-related equipment issues. This issue is lessening in areas such as physical ergonomics (e.g., where anthropometric data are often built into modern computer-aided design packages so that the different sizes and shapes of operators and maintenance workers can be considered). However, in more complex, or psychological, areas such as workload or being able to deal with multiple alarms, designers often have less assistance, and so need to rely more on trial and error or their own professional judgment.
3. Competing priorities. Designers generally work for larger equipment manufacturers, who sell equipment to mining companies. Consequently, there is a gap between the designer and the ultimate health, safety, or well-being of the end user. Where design compromises need to be made (e.g., lower price, high usability, or increased productivity of equipment), then the immediate demands of pleasing clients in terms of technical aspects may take precedence over "softer" and longer term aspects such as usability.
4. Professional pride of the designer and/or emotional investment in an idea. Designers may be reluctant to change something they have created, especially when human factors is not part of their core training.
5. Perceived costs. As seen earlier, human factors costs and benefits are difficult to fully quantify. For example, "adding" usability may be seen as an additional cost, especially if only considered later in the design process.

Better and more accessible human factors information (such as, we anticipate, in this book) can help with many of these. However, the issue of competing priorities is perhaps the most difficult to deal with. Rhetoric such as "Safety is the number one priority" and "Consider the end user" is certainly worthy, but practicalities can intrude: design milestones, long-standing design customs, and order books may have a significant influence. The mining human factors community needs to continually make the case that good human factors design is good design—and it can have safety, health, and productivity benefits.

2.9 Operability and maintainability analysis technique (OMAT)

Much of the material in this chapter so far has been quite theoretical. To conclude this chapter, a recent project by the University of Queensland for the Australian Coal Association Research Program (ACARP) to develop and evaluate OMAT will be presented (based on work by Horberry et al., 2009). This ties together and makes concrete many aspects of the equipment design process, safety in design, and using human factors information. It was directed at haul trucks used at surface mine sites, but could be applicable to the design of most mining equipment.

2.9.1 The importance of designing mobile equipment for maintainability and operability

To begin by formally defining the terms used in this section, *maintainability* is the ease with which equipment can be repaired safely in the least time, and *operability* is the ease with which equipment can be operated safely and in the optimal amount of time. As seen below, poor operability and maintainability of equipment can produce major safety and performance disbenefits. Therefore, it is vital that mobile equipment used at surface mine sites is both operable and maintainable.

Beginning by looking at figures from the United States that highlight the impact of mobile equipment upon safety, the National Institute for Occupational Safety and Health (NIOSH; 2009) reported that, from 2000 to 2004, the two leading causes of surface mining fatalities were as follows:

- 37 percent powered haulage
- 25 percent machinery

Large mobile equipment such as haul trucks have also been identified as a significant contributor to nonfatal lost-time injuries in surface mining; for example, they were an escalating causal factor to the overall breakdown event (Queensland Mines and Quarries, 2005). A *breakdown event* is defined as the point in which things start to go wrong, and ultimately lead to serious injury or disease. At the time of writing, the most recent Queensland, Australia, data available (2007–2008) support this, which again confirms the role of mobile equipment as a contributor in many lost-time injuries, high potential incidents, and disabling injuries (Queensland Mines and Quarries, 2008).

A comparable story is present for maintainability issues. Looking at more historical data, MSHA data for 1978–1988 suggests that maintenance accounted for 34 percent of all lost-time injuries (quoted by Horberry et al., 2009). Although old, this MSHA information is also consistent with more recent Australian data. An initial analysis of the 393 lost-time injuries recorded by the Queensland Government Department of Mines and Energy over the period from July 2002 to June 2004 indicated that 123, or approximately one-third, could be directly identified as being caused during maintenance activities (quoted by Horberry et al., 2009).

Besides deaths and lost-time injuries, the amount of avoidable downtime due to poor maintainability is more difficult to ascertain but nonetheless important. A British Coal report revealed that more man-shifts were devoted to maintenance operations than to coal production, and the proportion of time being devoted to maintenance is increasing (Mason and Rushworth, 1991). The repercussions of increased maintenance are increased labour costs, material costs, production disruptions, and probability of injury or illness. In the past, the design inadequacies related to maintenance tasks were typically related to poor access, inadequate provision of lifting points, and the need for excessive manual forces (Mason and Rushworth, 1991). However, as the design of earth-moving equipment is becoming more intricate and complex, the preventative and corrective maintenance tasks are increasing in occurrence, duration, and complexity, subsequently producing losses to coal production.

As a final piece of evidence about the importance of design, the Queensland Mines and Quarries Safety Performance and Health Report quantified the organisational causal factors associated with high potential incidents (HPIs) from January 2002 to June 2005. Design and maintenance management accounted for 11 percent and 12 percent of all HPIs, respectively (figures quoted by Horberry et al., 2009). It is unclear how much of the production loss in operating and maintaining large surface mining equipment is related to design issues, but the proportion is likely to be significant. This therefore suggests that significant safety benefits can be obtained through improved (and human-centred) equipment design.

In sum, the above outlines that many incidents and accidents are due to equipment design inadequacies, in either maintainability or operability, and are therefore theoretically preventable.

2.9.2 The beginning of the Earth Moving Equipment Safety Round Table (EMESRT)

Mining companies around the world have struggled with the need to ensure that earth-moving equipment is designed to be operated and maintained under all site conditions without causing harm to people. Designing haul trucks for safer operations and maintenance is therefore a major objective in managing a protected, efficient, and thereby competitive workplace.

With this in mind, from 2004 to 2006, a number of multi-national mining companies started discussing the concept of a joint customer approach to improve the human factors design of earth-moving equipment at the factory level. At the end of this period, a new approach to engaging with original equipment manufacturers (OEMs) was agreed: to define the "problem" rather than dictate the "solution." In other words, the industry risk management approach defined the basis of the equipment design risks, which went above and beyond the reliance purely on existing standards. This eventually resulted in a multicompany industry initiative known as EMESRT (Minerals Industry Safety and Health Centre, 2009).

EMESRT was formally established in 2006 by six of the major international mining companies. The purpose of this initiative was to establish a process of engagement between OEMs and mining customers, a process designed to accelerate the development and adoption of leading practise designs of earth-moving equipment to minimise risks to health and safety through a process of OEM and user engagement. EMESRT views operability and maintainability as major design challenges in large surface equipment.

As seen earlier in this chapter, considering the entire life cycle of an asset is a good engineering practise that results in improved safety, design, training, operation, modification, and maintenance. Identifying critical design characteristics that affect operability and maintainability and implementing those changes early in the design stage ensure the lowest cost and the greatest ease of implementation (Sammarco et al., 2001). This approach is within EMESRT's vision and would minimise the need for retrofit and corrective maintenance. As noted earlier in this chapter, many of the technical advances in equipment design over the past few decades have not been matched by advances in the human factors design of mining equipment. Local equipment dealers help to address this by often retrofitting equipment according to mining customer requirements to mitigate risks that have not been fully designed out by the OEM.

Some mining companies, individually, have attempted to establish a process for improved design that incorporates site usage of equipment into the design process. The time and effort required to achieve successful outcomes have weighed heavily on each company, which is evidence that other approaches, using carefully design methods, are needed. Clearly, then, a better safety in design approach that is propagated by EMESRT member companies and others could pay dividends.

It was envisioned that the implementation of a comprehensive equipment review process, including operability and maintainability, initially implemented at all design and build phases, initial site operation, and finally post retrofit, would help ensure that all critical design issues related to operability and maintainability tasks are addressed in the asset life cycle. The OMAT tool was developed to be such a risk assessment process, intended to deliver a piece of earth-moving equipment that will ensure the safety and well-being of all operators and maintainers.

2.9.3 Previous techniques for maintainability and operability assessment

Although this exact approach is new, it should be noted here that there has been proactive and systematic analysis of equipment maintainability and operability in the past. Indices for the mining industry have been generated in an attempt to mitigate the safety implications that can be controlled through proactive equipment design, most notably those included by the Bretby Indices by British Coal. The British Coal Technical Research and Services Executive (TRSE) produced many reports and papers (e.g., Mason and Rushworth, 1991) explaining and applying two methods called the Bretby Maintainability Index (BMI) and the Bretby Operability Index (BOI). The Bretby Operability Index was later revised to create an additional index, the Bretby Operability Retrofit Index (BORI).

The Bretby indices were, however, largely prescriptive, and specified and scored aspects of equipment against predefined elements. All machines have differing maintainability and operability priorities dictated by their potential consequence and occurrence. For example, sightline and visibility are more important to the safe operation of a haul truck compared to noise. Therefore, when designing for haul trucks, the visibility issues may be weighed heavier than those of noise.

The BMI predominately focused on ergonomic and human factors issues and disregarded safety implications from fire, dust, noise, working at heights, collision avoidance and detection, isolation, vibration, guarding, tyres and rims, and machine stability. Similar shortcomings apply to the BOI and BORI.

For example, fire and electrical factors account for a great majority of incidents in surface coal mining, specifically 33 percent in Queensland in 2005 (Queensland Mines and Quarries, 2005). The BMI has failed to include fire and electricity as critical design considerations and therefore risks the production of an unsafe machine by the OEMs and the potential acceptance of such a machine by the mine sites. The BOI and the BORI include a thermal environments category, yet the elements within that category are not representative of the main risks involving fire. The abovementioned critical causal factors, such as fire and electricity, were used in the OMAT process.

Finally, again, the Bretby indices are largely prescriptive. The dominant paradigm in safety management is now for a risk-based process, in part due to the inadequacies of the prescriptive approach in being able to prescribe for all situations, tasks, and equipment types.

2.9.4 *The scope of OMAT*

Firmly within the industry risk management approach (rather than being solution prescriptive), the OMAT tool is intended to help designers and mine site personnel identify, understand, and provide solutions to the risks people face when operating and maintaining new and current earth-moving equipment, specifically through the application of human factors information.

An iterative process was used to develop and evaluate OMAT that included reviewing related literature and techniques, creating an initial version of OMAT, performing a desktop review of the technique with industry personnel, running a first trial of the technique at an operational site (involving users from two major mining organisations, and an equipment manufacturer), running a second site and manufacturer trial of the technique, and making minor modifications as required.

Ideally, all newly purchased equipment should arrive at the mine site with all safety issues eliminated, and with no need for substitution controls, isolation or engineering controls, administrative controls (other than training in the operation of the equipment), or personal protective equipment. As far as possible, the risk should be designed out. However, as seen in the above incident statistics, this is not the current reality.

Acknowledging that there will be some residual safety concerns that have not been designed out when the equipment arrives at the mine site, it is imperative that the mine site also systematically analyses it. Also, all mine sites are unique and must often make some adjustment to their layout and facilities to safely accommodate new equipment, even if previously reviewed by the OEM. OMAT was therefore developed to be employed for mine site usage during operation or modification phases of equipment to address any residual risks, to examine site-specific risks related to newly purchased equipment, to investigate equipment-related incidents, or as a basis for new equipment purchase. In addition, current risk assessment processes used by mine sites do not usually supply their outcomes in such a way that the designers can readily understand and integrate them into new designs; OMAT was specifically created to address this issue.

The OMAT process has been aligned with existing OEM design processes (including major OEM design milestones, or "tollgates"). Most OEM equipment design processes currently rely heavily on failure modes effects analysis (FMEA), which is typically used to understand the mechanical and electrical failure modes of discrete components or systems. OMAT focuses specifically on human interaction with equipment, which is commonly disregarded in the FMEA. Performing both an FMEA and an OMAT would identify both the mechanical and human aspects in relation to the equipment.

OMAT has also been developed in alignment with the above-mentioned EMESRT scope, specifically by providing processes for user engagement and using structured human factors information to identify and assess the risks. Some key human factors areas in design that are considered are manual tasks, visibility, cognitive demands, controls and displays, reach and clearance, noise, and workspace layout. Also to be considered are the more safety-oriented equipment factors that can also contribute to unwanted events: fire, dust, confined spaces, electricity, and working at heights.

2.9.5 OMAT process

OMAT is a task-oriented risk assessment process that focuses on human factors risks related to mobile mining equipment design. Investigating such risks in all the operational or maintenance tasks involves six OMAT steps (after an initial stage O, to identify the need and establish the risk context,), with the first four being conducted in a joint OEM and mine site user workshop (see Figure 2.5).

In more detail, the six stages are as follows:

1. Based on a comprehensive list of all operations and maintenance tasks performed using the equipment, the OMAT process begins by prioritising the critical tasks (based on a set of safety and human factors criteria). Although ideally all tasks would be analysed, time restrictions may limit the remaining OMAT process to only focus on the highest priority ones.
2. Describing and analysing the constituent steps in these priority tasks. In effect, this is a task analysis, whereby each task step and its order are uncovered. Where deviations and/or shortcuts are possible, these should be noted.
3. Identifying risks at each of the task steps. Using the types of matrices commonly used in the mining industry (e.g., five-point severity and likelihood scales), the risks are identified, noting any current controls employed (e.g., guardrails for working-at-height tasks). The hierarchy of control is of assistance in this stage to analyse the risks.
4. Developing solutions. For the risks identified in step 3, solutions are developed. Again, time restrictions may restrict the focus to purely the highest priority risks. The solutions developed here should be primarily design solutions that are towards the top of the hierarchy of control.
5. Receiving feedback on the solutions from mine site users. The solutions developed in stage 4 would usually be further developed (e.g., in terms of technical specifications) by the OEMs. However, to continue the process of user-centred design, these further-developed solutions should then be evaluated by representative mine site users and any feedback noted.

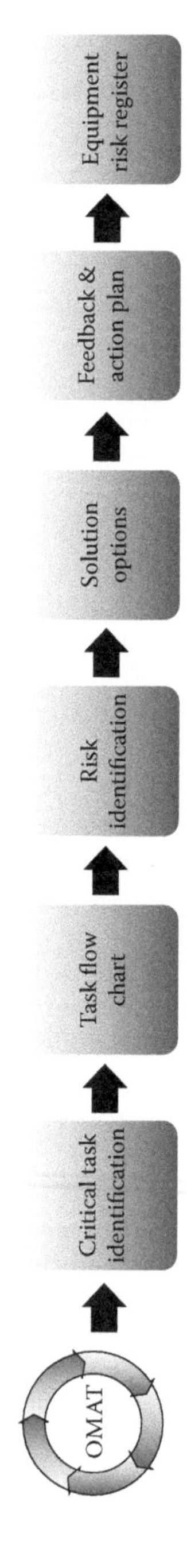

Figure 2.5 The OMAT process (Adapted from Horberry et al., 2009).

6. Maintaining a risk register. To keep track of the whole process, a risk register should be maintained. This would include who is responsible for which design aspects, and by when. This information may be especially useful for mine sites to allow them to develop controls for any remaining equipment-related risks.

A full description of the OMAT process and further EMESRT information can be found at www.mirmgate.com (MIRMgate, 2010).

chapter three

It is not just about design

Mining equipment operations and maintenance

3.1 Elements in the mining system

Everyone involved in mining, or the minerals industry more generally, is aware that it is a complex system where people, procedures, and equipment need to interact safely and efficiently in order to get the work done; however, to unpack this idea slightly, the system is characterised by the following:

- *A diverse group of people.* This includes differences in people's sizes and shapes, physical strengths, motivation, training and education, perceptual and cognitive styles, skills, expertise, and age. Many of these will be further reviewed in this book; however, the important point to make here is that there is no single "one-size-fits-all" categorisation to describe the people working in mining. As will be seen throughout this book, this has consequent implications for mining equipment design and operation.
- *Diversity of company cultures.* Many mining companies are multinational, and own sites in countries with diverse cultures; however, each also has its own unique organisational structural and cultural differences. They differ on many important dimensions such as their safety management systems, recruitment policies, and corporate leadership styles. Just as there is no typical mining employee, there is also no "typical" mining company.
- *Wide variety of national laws, regulations, and guidelines.* Given the above-mentioned multinational nature of mining, different countries have differing laws, regulations, political pressures, standards, and guidelines. For example, safety regulation varies immensely around the world, and particularly in the responsibility given to companies to individually manage risks versus comply with regulation (e.g., Poplin et al., 2008), and the onus placed on equipment manufacturers for equipment safety. Indeed, regulation can be quite

different between different states in the same country (e.g., in the United States or Australia).

- *Different procedures, rules, practises, and cultures at mine sites.* Cultures, practises, formal and informal rules, and procedures vary within companies, as well as across them. For example, mines in a small geographic area, such as the coal-mining region of the Hunter Valley in Australia, often have different procedures, practises, and cultures, stemming in large part from their history. And when it is considered that some mines have a history of ownership by different companies, it is not too surprising to see that the mine culture may be a stronger influence on site practises than the culture of the owner company of the day. Differences relevant to equipment design might include policies for the separation of people and mobile equipment (e.g., the exact "give-way" rules), teamworking procedures, or safety communication methods, together with less obvious cultural differences such as attitudes towards risk taking.
- *Many equipment manufacturers and suppliers.* Added to the mix is the variety of different equipment manufacturers, dealers, after-market suppliers, technology developers, and so on. For example, looking at surface mining, there are five or six major manufacturers of large mobile equipment (such as haul trucks). Although they usually follow international standards for equipment design (where they exist), the differences in the equipment for maintenance or operational purposes can be quite significant (as will be seen in Chapter 8, which examines controls and displays).
- *Variation in the built environment.* Of course, mining methods differ; at the broadest level, of course, they differ between underground and surface mining, but also differences exist as a function of the substances being mined or processed. Coal mines in particular have additional complexity associated with the additional environmental hazards involved with this. But, going into more detail, the design of the built environment between mines using the same mining method can significantly differ. This includes differences in issues such as access to the mine site or roadway construction. Similarly, a mine that is built near a major population centre has a very different set of requirements and needs from one in a remote location. Whereas a mine near a major population centre usually has better access to a source of personnel and might be better equipped to transport goods (e.g., fuel or ore) to and from the site, it may have greater issues with respect to site security and/or community relations.
- *Many uncertainties in the natural environment.* Given the worldwide nature of mining, the natural environment can certainly have a major influence. The variation of temperatures between sites in the tropics compared to those in the polar region can clearly be huge.

> Other key issues in the natural environment include humidity levels, dust, darkness periods, and rain, snow, and wind levels. All can have a huge impact upon how equipment is set up, serviced, and operated. One example is from surface mining in the Arctic region (e.g., Canada, Alaska, or Russia), where the low position of the sun on the horizon at certain times of the year can cause significant visibility issues compared to in other more temperate locations: fitting of highly adjustable sun visors (by the manufacturers, the dealers, or more likely the site itself) is essential.

The complexity of the elements in the mining system described above within which equipment is operated and maintained points to the observation that whilst the consideration of human factors in the design of equipment is essential, it is not sufficient. Consideration of how the equipment is maintained, set up, and used at a site is also clearly of key importance to both safety and performance. The remainder of this chapter will expand on this theme. To illustrate the points, it will focus primarily on mobile equipment operations. Similarly, it will focus here more on safety than on performance. However, the principles are applicable more broadly.

3.2 *Safety in the operation of mobile equipment*

The issues will be illustrated by means of work done by one of the authors examining forklift trucks and similar mobile equipment (Horberry et al., 2004; 2006). This was chosen because it demonstrates many key points about design, maintenance, and operation. Also, forklift trucks are commonly used worldwide in many areas of mining, as well as in other industrial and manufacturing operations. They have become essential for materials handling (Feare, 1999), as well as a flexible vehicle for many freight movement tasks in mining and elsewhere. Whilst they offer many benefits, such as improving work efficiency or reducing manual handling, they can also pose a major occupational hazard, especially where used in close proximity to "pedestrian" workers at a mine site. Crashes between these vehicles and unprotected workers frequently cause severe injury or even death. For example, in the United States there were nearly one hundred forklift fatalities per year (Bureau of Labor Statistics, 1998) in the 1990s. Similar figures are found for countries such as Canada and the United Kingdom (Horberry et al., 2004). Forklift crashes also produce a large number of serious injuries.

Due to the design and operational requirements of these vehicles, certain risk factors are almost inherent. The versatility, ease of operation, power, and flexibility of the vehicles that enable them to be used for a multitude of tasks in a wide variety of mining environments constitute some of the major risk factors. Vehicle stability is a primary safety concern, often due to a forklift's narrow track and high centre of gravity. Speed is also a determinant in many forklift accidents, for both the drivers and pedestrian workers; their rate of travel has a direct bearing on the level of risk to which pedestrian workers are exposed (Larsson and Rechnitzer, 1994).

So, as with much mobile mining equipment, there are many benefits from forklift use, but also some serious safety issues. It should be noted that some mining organisations still have ill-defined procedures and traffic management policies with respect to such truck operations. Controls often focus on the lower areas of the hierarchy of control (as defined in Chapter 2), typically operator protection. A lack of consideration of the interface between pedestrian workers and mobile equipment (both in mining and elsewhere) has resulted in many of the more serious injuries and fatalities (Larsson and Rechnitzer, 1994).

A collection of engineering and administrative controls is therefore required; this often should include improved traffic engineering, safer initial design, and, where necessary, the use of retrofitted vehicle technologies such as speed-limiting systems or seat belt interlocks to reduce risks to operators and pedestrian workers (Horberry et al., 2006). Where such a broad array of interventions was applied in a case study in the manufacturing environment (in many ways similar to mining), the results were positive: less critical interactions occurred between mobile equipment and pedestrians or other vehicles, and the implementation of the speed limiters and seat belt devices was generally deemed to be acceptable to the operators, providing that the work process was not overly slowed (Horberry et al., 2004).

3.3 *Different types of factors involved in mobile equipment incidents*

The solutions to prevent or mitigate mobile equipment accidents extend beyond purely the original design of the equipment. At the broadest level this is nothing new; the work of William Haddon in the 1970s pointed to similar things (albeit in a different domain). Haddon (1972) devised a matrix of broad categories of factors and phases of injury; this was not purely for mobile equipment incidents in mining, but certainly is applicable here. This matrix has been widely applied in injury prevention to show how different factors contribute to different incident phases. In this vein, Horberry et al. (2006) created an example "Haddon matrix" for industrial mobile equipment incidents and accidents and possible countermeasures. An adaptation of this for mining equipment is shown in Table 3.1. The factors shown in italics are considered to be design-rated factors, and non-italicised factors are thus controls and countermeasures not directly design related (although, of course, significant grey areas exist). This therefore shows the scope of different controls, and why issues wider than

Table 3.1 A Haddon Matrix for Mining Equipment

Phase	Factors			
	Host (mobile equipment operators and pedestrians)	Vehicle (forklift or other mobile equipment)	Physical environment (mine site)	Social environment (company policies and rules)
Pre-event	*Driver vision* *Pedestrian visibility* Alcohol use *Fatigue* *Equipment use training* *Hearing and noise* Use of pedestrian walkways Obeying exclusion zones and crossings Obeying signage Safety inductions	*Brakes* *Tyres* *Load characteristics* *Speed of travel* *Turning radius* *Direction of travel* *Tyne position* *Tyne tilt and angle* *Ease of control* *Visibility from cabin* *In-vehicle warning devices*	Site design Visibility of hazards Surface friction Uneven ground Blind corners Intersection Crossings Height and width of doorways Signage Signals Exclusion zones Pedestrian walkways	Overall attitudes about safety Attitudes about alcohol Rostering Logistics planning *Maintenance scheduling* *Training* Speed limits and enforcement
Event (accident or incident)	*Safety belt use* *Use of personal protective equipment (PPE)* Emergency manoeuvring skills	*Stability control* *Vehicle size* *Characteristics of contact surfaces* *Load containment* *Rollover protection structures (ROPS)* *Seat belts and seat bars*	Guardrails Speed limits Characteristics of fixed structures	Attitude about seat belt use Attitude about PPE use
Post event	Operator age Physical condition	*Fuel cut-off* *Deadman's control*	*Emergency communication systems* Easy access to emergency medicine	First aid training Support for trauma care and rehabilitation

Source: Adapted from Horberry et al. (2006).

vehicle design are important. Also, whilst the list is in no way exhaustive, it shows how different countermeasures and controls can be used for different phases of the event (not just to prevent incidents per se).

3.4 Haddon's countermeasure principles

Of course, as seen in Chapter 2 regarding the hierarchy of control, without eliminating all vehicles or at least their energy source, no single control measure can prevent all mobile mining equipment incidents and accidents. However, principles to guide the formation of other countermeasures do exist. The hierarchy of control is one set of principles, but it is more of a broad overview. Another influential set, which extends in more detail into operations and maintenance, was devised by Haddon (1973). He defined ten logically distinct technical strategies for injury control. These are shown below. To illustrate them, they continue the example of mobile mining equipment incidents and accidents (adapted from Horberry et al., 2006, who applied these principles to help identify and prioritise technology-based safety countermeasures for trucks). It is argued that these can help guide the identification and assessment of risks, and provide a structured way to develop controls.

3.4.1 Principle 1: Prevent the creation of the hazard

The obvious measure here would be to eliminate mobile equipment altogether (as also shown at the "top" of the hierarchy of control). Where this is not possible or practical, or where the hazards are more specific, examples of applications of this principle might include the following:

- Prevent overloads (to prevent issues associated with equipment being overloaded, such as increased risk of rollover, poor vehicle handling, restricted visibility, or spillages).
- Equipment maintenance (to reduce the risk of equipment malfunction).
- Eliminate the need to reverse (to remove the risk of reversing incidents).

3.4.2 Principle 2: Reduce the amount of the hazard

The "amount" can be defined in several ways. Examples might include the following:

- Speed limiting.
- Limit number of trucks and their size.
- Minimise the number of trips.
- Reduce reversing.

3.4.3 Principle 3: Prevent the release of the hazard

As with Principle 2, several ways are possible. Examples here might include the following:

- Operator or maintainer vision testing.
- Access control (to prevent unauthorised use of the vehicle).
- Stability and tilt controls.
- Load stability warnings.
- Maintain level roadways.
- Match equipment to accessways and openings.

3.4.4 Principle 4: Modify the rate of release of the hazard from its source

In the example here, this might largely focus on vehicle safety systems. Examples might include the following:

- Automatic mobile equipment braking
- Proximity sensors (see Figure 3.1)
- Vehicle airbags

Figure 3.1 Computer-generated example of proximity sensors for mobile equipment.

3.4.5 Principle 5: Separate the hazard from that which is to be protected in time and space

Examples of separation might include the following (see Figure 3.2):

- Exclusion zones
- Overhead pedestrian bridges
- Pedestrian walkways
- Traffic lights

3.4.6 Principle 6: Separate the hazard from that which is to be protected by a physical barrier

This principle overlaps with Principle 5. Examples might include the following:

- Some types of personal protective equipment (PPE) use
- Guardrails or barriers around walkways (also as shown in Figure 3.2)

3.4.7 Principle 7: Modify relevant basic qualities of the hazard

An example might include soft or crash-absorbent covers on protruding parts of the truck.

Figure 3.2 Pedestrian walkway separation from mobile equipment.

3.4.8 Principle 8: Make what is to be protected more resistant to damage from the hazard

Focusing on pedestrian workers, an example might be PPE use.

3.4.9 Principle 9: Begin to counter damage done by the hazard

Examples here might include the following:

- Impact sensors and notification
- Rapid emergency service access
- First aid provision

3.4.10 Principle 10: Stabilise, repair, and rehabilitate the object of damage

In the mobile equipment and pedestrian worker case, examples might include the following:

- Rehabilitate the operator.
- Repair the truck.
- Equipment remedial maintenance.

Clearly, human factors knowledge comes into play in the identification, development, and implementation of controls resulting from many of these ten principles (for example, the design and implementation of PPE, walkways, or vehicle access controls).

3.5 Conclusions

The last two chapters have shown in general terms the importance of design, site operations, and maintenance upon the safety and efficiency of mining equipment. Key concepts such as the hierarchy of control have been introduced, and human-related issues have been shown to be of central importance in the whole area. The focus of this book will now switch, and it will begin to examine specific human factors issues with mining equipment. This process begins by first considering manual tasks in mining equipment operations and maintenance.

chapter four

Manual tasks

4.1 Introduction

Manual tasks are any tasks performed by people that involve the use of force to lift, push, pull, carry, move, manipulate, hold, or restrain a person, body part, object, or tool. Operating and maintaining mining equipment inevitably, and inescapably, involve manual tasks. Manual tasks include activities such as driving, manipulating controls, handling supplies, lifting access hatches, changing fittings, using rattle guns, and many others (see Figure 4.1A–F). Whilst perhaps not as hazardous as some others traditionally encountered in mining, even maintaining a static seated posture whilst viewing a display and manipulating a joystick is still a manual task.

Many manual tasks have the potential to cause injury. Injuries occur when forces on parts of the body are, either instantly or over time, greater than the body part can withstand. Tissues at risk of damage due to manual tasks include bones, muscles, tendons, ligaments, articular cartilage, nerves, and blood vessels. Table 4.1 provides examples of typical injury narratives describing manual task–related injuries associated with a piece of mining equipment (an underground coal continuous-mining machine). It is clear from these narratives that both operation and maintenance are involved. What is less clear from these injury narratives is that the likelihood of an injury occurring may increase as a consequence of an accumulation of damage over time, and this damage may be caused by tasks other than the one being performed at the time the injury is noted and reported.

Optimal design of mining equipment has the potential to reduce injury risks by ensuring that the manual tasks involved in the operation and maintenance of equipment involve minimal injury risks. Of course, ensuring that the operation and maintenance of equipment can be achieved without requiring the performance of potentially hazardous manual tasks will also have advantages in the form of reduced fatigue, fewer operator errors, and reduced maintenance time.

To appreciate the design goals in this area, it is helpful to consider the sources and nature of injury risks associated with manual tasks. The mechanisms of injury to specific tissues vary; however, injuries associated with manual tasks may be generally characterised as having either sudden or gradual onset. Sudden-onset injuries are associated with

(A)

(B)

Figure 4.1 (A–F) Examples of manual tasks in mining.

(C)

(D)

Figure 4.1 (Continued) (A–F) Examples of manual tasks in mining.

(E)

(F)

Figure 4.1 (Continued) (A–F) Examples of manual tasks in mining.

Table 4.1 Manual-Task Injury Narratives

Whilst flitting a continuous miner, he bent down to lift a continuous miner cable over his head onto a cable roller, straining his lower back.
Whilst lifting a roof mesh onto the top of a continuous miner, he strained his neck and left shoulder.
Whilst lifting a hydraulic jack under the head of a continuous miner, he strained his groin.
Whilst assisting with boom repair on a continuous miner when holding the weight of a 15 kg large steel pin, he injured his lower back.
Whilst attempting to lift a temporary roof support cylinder with another person back onto a continuous miner, he felt lower back pain.

a relatively short exposure to forces which exceed a tissue's capability. In contrast, gradual-onset injuries occur as a consequence of relatively long-term exposure to forces. In the latter case, the general mechanism of injury is believed to be an accumulation of microdamage which exceeds the tissue's capacity for repair. Injuries may also occur as a combination of both general mechanisms, where a history of cumulative loading leads to reduced tissue tolerance which is then exceeded by short-term exposure to a relatively high force.

Considerable attention has been paid to the associations between various possible risk factors and occupational musculoskeletal injuries. A relationship exists between musculoskeletal disorders and prolonged exposure to forceful exertions, awkward and static postures, and repetition (National Institute for Occupational Safety and Health [NIOSH], 1997), and these factors are considered to be direct risk factors. Injuries are particularly associated with exposure to multiple direct risk factors. Indirect risk factors include environmental conditions (e.g., heat and cold) and psychosocial issues (e.g., stress, work pacing, lack of control, and conflict with peers or supervisors).

4.2 Direct manual-task risk factors

4.2.1 Force and speed

An important factor in determining the likelihood of injury to a specific body part is how much force is involved. Historically, the mass of objects being handled has been the main focus of attention; however, the force involved in a task depends on a number of other factors as well. As examples: in lifting and lowering tasks, the force required by the back muscles depends as much on the distance of the load from the body as it does on the mass of the load; and the shoulder torque required to position a drill steel or bolt is a function of both the mass of the steel, or the bolt, and the distance it is held from the shoulder. Similarly, if the task involves pushing or pulling the force involved will depend on the frictional properties of the situation as well as the mass of the object being handled. Other manual tasks may not involve the manipulation of any load; however, high forces can still be required.

If the force which is required to be exerted by a body part is close to the maximum the person is capable of, then the risk of sudden injury is high, and urgent redesign is indicated. Even if the forces involved are not close to maximum, the task may pose a high risk of injury if the body part is also exposed to other risk factors. High-speed movements (e.g., hammering and throwing) are an indication of elevated risk, mostly because high speed implies high acceleration, which in turn implies high force, especially if the speed is achieved in a short time. Such "jerky" movements are a sure indication of high exertion of the body parts involved. This also includes rapid changes in the direction of movement. The strength of muscles is in part dependent on the speed at which they shorten, and high-speed movements consequently reduce the strength of the muscles producing the movement. Another situation occurs when high-impact force is applied by the hand to strike an object or surface; in this case, there is a resultant high force applied to the hand by the object or surface being struck when contact is made.

The magnitude of the force relative to the capabilities of the body part is crucial in determining injury risks. For example, the small muscles of the hand and forearm may be injured by relatively small forces, especially if the task also involves extremes of the range of movement at a joint, whilst larger forces may comfortably be exerted by the lower limbs. Guidance regarding appropriate force levels to be applied to different controls is available (e.g., US Department of Defense, 1999), although care is required to ensure that the values are appropriate for the user population, a topic which is examined in more detail in Chapter 5.

4.2.2 Body posture

The postures adopted during a task influence the likelihood of injury in a number of ways. If joints are exposed to postures which involve extremes of the range of movement, the tissues around the joint are stretched and the risk of injury is increased. Ligaments, in particular, are stretched in extreme postures. If the exposure to extreme postures is prolonged, the ligaments do not immediately return to their resting length afterwards. Tissue compression may also occur as a consequence of extreme postures, for example, extreme postures at the wrist increase the pressure on the nerve which passes through the carpal tunnel.

The strength of muscles is also influenced by the posture of the joints over which they cross. Muscles are weaker if they are shortened, and this effect will be greatest when the joints approach the extreme of the range of movement. Consequently, one general principle for the design of equipment is to ensure that the operation and maintenance of the equipment can be achieved by all operators without requiring postures which involve extremes of the range of movement at any joint.

Some non-extreme joint postures are also known to be associated with increased risk of discomfort and injury. These include trunk rotation, lateral trunk flexion, and trunk extension; neck extension, lateral flexion, and rotation; and wrist extension and ulnar deviation. Some other postures increase the risk of injury without involving extremes of the range of movement. These can be called *awkward postures*, and can be defined as any posture which causes discomfort. Such postures can occur without significant deviation of the joint from neutral, especially if the orientation of the body with respect to gravity is altered. Adoption of awkward postures during the operation, and especially maintenance, of equipment is frequently associated with either restricted visibility or restricted access. Optimal equipment design will ensure that the manual tasks required for the operation and maintenance of equipment involve movements within a normal range about neutral.

4.2.3 Movement and repetition

The performance of manual tasks which involve low to moderate exertion, slow to moderately paced movements, no awkward postures, and varied patterns of movement is healthy. Little or no movement at a body part elevates the risk of discomfort and injury because the flow of blood through muscles to provide energy and remove wastage depends on movement. Tasks which involve static postures quickly lead to discomfort, especially if combined with exposure to other risk factors.

If the task involves repetitively performing identical patterns of movement, and especially if the cycle time of the repeated movement is short, then the same tissues are being loaded in the same way with little opportunity for recovery. Such repetitive tasks are likely to pose a high risk of cumulative injury if combined with moderate to high forces (or speeds), awkward postures, and/or long durations.

4.2.4 Duration

If a task is performed continuously without a break for a long time, the tissues involved do not have opportunity for recovery, and cumulative injury risk increases. This is especially likely if the task involves a combination of moderate force, little or repetitive movement, and awkward postures. Changing tasks can provide recovery if the second task involves different body parts and movement patterns. The appropriate task duration also depends on environmental factors such as vibration.

4.3 *Assessing manual-task injury risks*

As a consequence of the complexity of injury mechanisms involved, the assessment of manual-task injury risks requires specialist tools, rather than the typical 5 × 5 workplace risk assessment and control (WRAC)–style risk assessment matrix. A model, "Procedure for Managing Injury Risks Associated with Manual Tasks," is available (Burgess-Limerick, 2008) which includes a manual-task risk assessment matrix incorporating consideration of both direct and indirect risk factors. A recent NIOSH publication (Torma-Krajewski et al., 2009) also provides a similar risk assessment strategy.

Part of the equipment design process should include identifying the potentially hazardous manual tasks involved in the operation and maintenance of equipment, and documenting the risks associated with these tasks. A paper-based system may be adequate; however, a software database (ErgoAnalyst; see, e.g., Figure 4.2) is available (see www.ergoenterprises.com.au). Where significant manual-task risks are identified, additional design control measures should be explored, ideally using measures that are high up the "hierarchy of control" (mentioned in Chapter 2), for example, elimination or substitution controls.

Where the elimination of injury risks through the manufacturer's design process is not possible, the documented manual-task risks

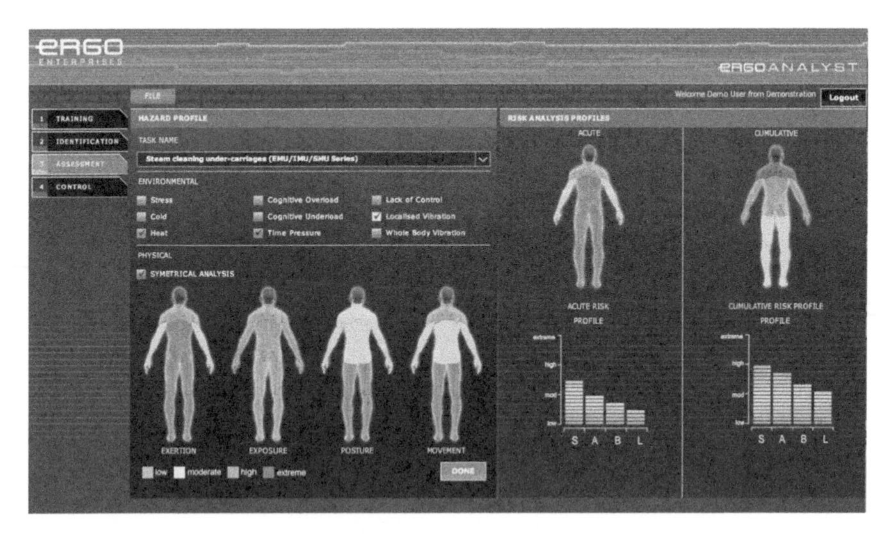

Figure 4.2 The ErgoAnalyst software.

associated with operation and maintenance should be considered within operational risk assessments prior to the equipment's acceptance. The operability and maintainability analysis technique (OMAT) process described in Chapter 2 could be of significant assistance here, for manual-task risks as well as other operational- and maintenance-task risks. Such a process is consistent with guidelines such as MDG15 (New South Wales Department of Primary Industries [NSW DPI], 2002), clause 3.6, which stipulates,

> *A suitably qualified person should review the ergonomic aspects of the equipment to ensure compliance with good practice. A report should be prepared by this person and supplied to the Operator before delivery.*

It is also consistent with MDG35.1 (Industry & Investment NSW, 2010), clause 3.2.2.5, which states,

> *The designer shall consider safety related aspects of ergonomic issues for persons carrying out – a) repetitive work when addressing the layout of all bolting plant components and their use; and b) maintenance work when addressing the layout of all bolting plant components and their use. An ergonomic assessment on the layout of all operator controls should be carried out. The assessment should be carried out by a suitably competent person.*

Obligation holders at the site level then have a responsibility to ensure that other design and administrative controls are identified, implemented, and monitored where required.

4.4 The place of "training" in manual-task injury risk management

Training is an important administrative control regardless of which design controls are employed, in that training in the appropriate way of performing work and using equipment should be documented by manufacturers and provided by employers. Implementing an effective manual-task risk management program also requires that staff are able to identify hazardous manual tasks, and are aware of the aspects of manual tasks that increase injury risks. This might legitimately extend to principles such as "Keep the load close" and "Avoid twisting." However, the evidence is clear that training in "correct" load-handling techniques is not effective in reducing injuries associated with manual tasks and, on its own, is not a satisfactory risk control strategy (e.g., Daltroy et al., 1997; Silverstein and Clark, 2004; Haslam et al., 2007; Martimo et al., 2007).

4.5 Conclusion

The operation and maintenance of mining equipment require miners to undertake manual tasks. The challenge for employers and manufacturers is to identify potentially hazardous manual tasks, involved in equipment operation and maintenance; assess these risks in a systematic way, taking into account the known direct risk factors of high exertion, awkward postures, frequent repetition of similar movements, and duration of exposure; and redesign the equipment to either eliminate the manual tasks or reduce as far as possible the exposure to direct risk factors. Administrative controls will be necessary to control the residual risks associated with manual tasks involved in both the operation and maintenance of mining equipment; however, training alone is not a satisfactory risk control strategy.

chapter five

Workstation design and anthropometric variability

5.1 Workstation design: Overview

At a most basic level, one aim of considering human factors in the design of equipment is to ensure that this equipment 'fits' the operators. This aim is complicated by the fact that potential operators vary considerably in size, shape, and physical abilities. The study of the physical dimensions of people, and particularly the variability in these dimensions, is called *anthropometry*. This chapter first addresses how the design of equipment should take this anthropometric variability into account, and then provides general principles associated with other common aspects of mining equipment, including access and egress, seating, and visibility.

5.2 Incorporating anthropometric data in workstation design

The use of anthropometric data in the design of equipment is an important component of human factors.

5.2.1 Types of anthropometric data

The most commonly available data are static one-dimensional measurements such as height, weight, circumferences, and lengths. Whilst these measurements may be useful to answer some simple questions related to equipment design, in general, they are not sufficient. For example, knowledge of one-dimensional arm length does not allow prediction of functionally relevant information such as reach distances, because these functionally relevant dimensions are a product of multiple dimensions. Static two-dimensional (e.g., body silhouette) or three-dimensional (e.g., surface scan) data may also be measured, but have similar limitations.

Dynamic anthropometric data such as workspace envelopes, reach distances, and force measurements are potentially more useful, but highly specific. For example, the force which can be exerted by any part of the body is highly dependent on the direction and speed of the movement, and the posture adopted to perform the test.

Anthropometric data gathered from a population are typically expressed in terms of percentiles, where the 50th percentile is the median (middle) value of the measurement obtained from the population, the 5th percentile is the value below which 5 percent of the population measurements fell, and the 95th percentile is the value below which 95 percent of the population measurements fell. It is important to remember that these values are only accurate for the population from which the measurements were made (the size and shape of populations drawn from different countries, and different occupations, may vary widely, as do the measurements of males and females and even the same people at different ages), and for the time at which the measurements were made (the size and shape of populations have steadily changed over the years); indeed, they strictly assume random sampling from the population, a condition which is seldom satisfied.

5.2.2 Sources of anthropometric data

Although not perfect, there are a reasonable range of anthropometric data available that, with careful use, are suitable for mining equipment design and evaluation. Data are available in anthropometry software packages such as PeopleSize (2008) and human factors textbooks specifically devoted to anthropometrics (such as *Bodyspace: Anthropometry, Ergonomics and the Design of the Work*; Pheasant and Haslegrave, 2006). Other sources include the following:

- AS4024.1704—2006 (Standards Australia, 2006) provides limited static one-dimensional data derived from "European surveys," with a focus on those data which may be necessary to design access (part 1703) and guarding (1704).
- The US Federal Aviation Administration (FAA) *Human Factors Design Standard* (HFDS; FAA, 2009), Chapter 14, presents static and dynamic anthropometric data from FAA technical operations personnel.
- The 540-page US Department of Defense *Military Handbook: Anthropometry of US Military Personnel* (DOD-HDBK-743A; US Department of Defense, 1991) provides detailed data (203 measurements) for US military populations derived from a variety of different surveys.
- The Man-Systems Integration Standards, revision B, provided by NASA (1995) provides 5th and 95th percentiles for static anthropometric measurements corresponding to a forty-year-old American

male and a forty-year-old Japanese female. Some functional anthropometric measures including strength are also provided.

- The Civilian American and European Surface Anthropometry Research Project (CAESAR; Society of Automotive Engineers International, 2010), a commercially available database of US, Canadian, and European civilians, includes both one-dimensional measurements and three-dimensional surface scans.
- Strength data from UK civilians for diverse ages are available from the UK Department of Trade and Industry (2000 & 2002; also see Peebles and Norris, 2003).

5.2.3 *Use of anthropometric data in design*

As difficult as it can be, obtaining relevant anthropometric data which describe the population of interest is just the start, and making use of the data is a non-trivial problem. The FAA HFDS (FAA, 2009) provides guidance in the use of the anthropometric data through a "design limits approach." This approach involves (1) selecting the correct measurement, (2) selecting the appropriate population data, (3) determining the appropriate percentile, (4) determining the corresponding measurement value, and (5) incorporating this value as a criterion for the design dimension. This "designing for extremes" approach represents one method of using anthropometric data which is suitable when either the largest or smallest potential user should be accommodated, and when to do so will not inconvenience others in the population. Clearances are one example where designing for extremes may be appropriate (in this case, for the largest user).

However, designing for the extremes is not appropriate where this will create inconvenience for other users (e.g., door handle height). In this case the designer may "design for the average," particularly where the adverse consequences for the extremes are not serious. If so, then designing for adjustability to accommodate both extremes may be required (e.g., adjustable seat heights). However, this approach is only beneficial if correct adjustments are actually made.

In all cases, it is likely that there will be some proportion of the population which any piece of equipment is not designed to accommodate. For example, it is unlikely that sufficient range of seat height adjustability can be provided in an underground coal-mining vehicle to accommodate 100 percent of the population. The restricted seam height is likely to ensure that the tallest potential operators will not have sufficient head clearance to the underside of the canopy. In this situation, there is an obligation for the designer to identify and communicate the limitations to the purchaser of the equipment, and for the employer to ensure these limitations are observed.

5.2.4 Issues with the use of percentiles: The myth of the 50th percentile person

Whilst one-dimensional anthropometric data may have utility in some situations, when multiple dimensions are involved, the design task becomes more difficult again. The problem is that there is not a tight relationship between different measures, and when employed for multiple dimensions the use of percentiles will accommodate fewer people than the percentiles suggest. To put it another way, there is no person who is 50th percentile on all measures. Daniels (1952) demonstrated that of 4,063 men, none were within the middle one-third for all of fifteen measurements relevant to clothing design. Similarly, a person who is 5th percentile on one measure (e.g., height) is unlikely to be exactly 5th percentile on any other measure.

The consequence of this is that when there are multiple dimensions to be considered, it is not sensible to attempt to design equipment to suit fictitious persons who are 5th or 95th percentile on all the relevant dimensions. Indeed, Robinette and Hudson (2006) documented an example where an aircraft cockpit which was so designed was found not to be suitable for a real 5th-percentile-height woman whose thigh size was at the 50th percentile (when the seat was adjusted high enough for the person to see, and reach the controls, the position of her thighs was impinged by the yoke controls). The cockpit was subsequently found to accommodate only 10 percent of the female population.

The alternative to using percentiles is to design for a range of real cases, chosen to represent the population of interest. Boundary cases in particular, which represent the extremes of combinations of two critical dimensions (e.g., sitting height and buttock–knee length), may be useful for design problems where the issues can be reduced to two dimensions. Dainoff et al. (2004) and Robinette and Hudson (2006) describe these methods in more detail.

5.3 General principles of workstation design

The following issues require consideration during the design and commissioning of mining equipment in general.

5.3.1 Clearance requirements

Adequate space (head room, knee room, and elbow room) is required to accommodate the largest potential users of equipment. Where appropriate, these specifications should include protective clothing, self-rescuer, battery, and any other equipment carried by the operator or maintainer. For vehicles, head clearance whilst driving is a key concern, particularly in underground vehicles. Figure 5.1 illustrates the difficulties encountered

Figure 5.1 Restricted headroom in an underground coal-bolting machine.

by a tall miner standing on the platform of an underground coal-bolting machine. The restricted space available makes the adoption of awkward postures a necessity to operate the equipment. Figure 5.2 similarly illustrates the confined space available in the cab of many underground vehicles. Note in particular that the miner's helmet protrudes above the rollover protection provided.

5.3.2 Access and egress, and fall prevention during operation and maintenance

Access to and egress from mining equipment are important, and sometimes overlooked, aspects of equipment design. Poor access can result

Figure 5.2 Confined cab space.

in musculoskeletal injuries due to overexertion and awkward postures, as well as increase the probability of slips and falls during access or egress. Injury records clearly demonstrate this is a common injury mechanism (Burgess-Limerick and Steiner, 2006, 2007). Another aspect of the design is to ensure that rapid egress is possible in emergency situations. Clearances for both primary access and egress, as well as emergency egress, should be adequate for the largest potential operator and include consideration of clothing, and self-rescuer and battery as appropriate.

Access and egress also require consideration of the step height and reach distance capabilities of the smallest potential operator. Standards such as AS3868 (Standards Australia, 1991) and AS1657 (Standards Australia, 1992) provide guidance for access and egress, and specify for example the maximum step height of 400 mm. Whilst the provision of ladders for access is common, the risk of falling is relatively high, and stairs are preferable.

For example, whilst the access system illustrated in Figure 5.3 is typical, an innovative access system incorporating stairs rather than ladder access has been provided in the underground dozer illustrated in Figure 5.4.

Figure 5.3 Typical heavy vehicle access.

Similarly, cut-out footholds are common, and whilst satisfactory for access are difficult to use during egress, as illustrated in Figure 5.5.

Access systems provided on haul trucks, such as that illustrated in Figure 5.6, and other large surface-mining equipment have historically been unsatisfactory, and this created a need for after-market access solutions provided by other vendors (e.g., Safe-Away, n.d.) to be fitted.

This issue was one of those identified by the original Earth Moving Equipment Safety Round Table (EMESRT) group, and was addressed by the Equipment Access and Egress design philosophy. This design philosophy aims to ensure that (in this case, surface) equipment is designed to include the following:

Adequate and suitable stairways, walkways, access platforms, railings, step and grab handle combinations, and boarding facilities, including an alternate path for disembarking in case of emergency.

Specific to hauling trucks, a priority outcome would also be ground entry to access on the driver's side, with the opportunity to locate isolation

Figure 5.4 "Industrea" dozer access.

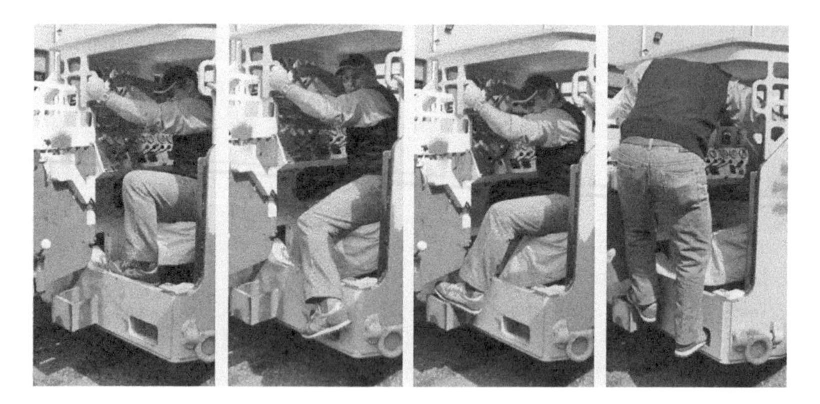

Figure 5.5 Cut-out foothold.

Figure 5.6 Haul truck access.

and other service points (hydraulic, air) near the driver's side operator access (EMESRT, n.d., www.mirmgate.com.au).

More recently—and, in part at least, encouraged by the activities of EMESRT—the original equipment manufacturers of haul trucks have provided much improved access systems.

Access and egress to equipment comprise a specific example of a more general design goal of reducing the possibility of slips and falls during operation, and ensuring safe access and working conditions during maintenance activities. This applies to fixed plant (e.g., Figure 5.7) as well as mobile equipment. The design of working and access platforms on equipment requires careful consideration, regardless of the height of the platform (e.g., Figure 5.8), as does the design of access systems for maintenance (e.g., Figure 5.9), and routine maintenance in particular (e.g., Figure 5.10).

5.3.3 Location and arrangement of workstation controls and displays

Human factors issues associated with the design of controls and displays will be dealt with in a separate chapter (Chapter 8); however, the location and arrangement of controls and displays form a critical part of the overall workstation design, and this aspect will be considered here.

The following (and, sadly, potentially conflicting) principles should be considered when determining, modifying, or evaluating the placement of equipment controls and displays:

Figure 5.7 Fixed plant.

Figure 5.8 Continuous miner access: Handrails.

1. *Avoid awkward postures*: All controls should be able to be operated without requiring awkward postures to be adopted. At the very least, all controls should be located within reach of the smallest potential user. The Man-Systems Integration Standards provided by NASA (1995, section 3.3.3.3.1, "Functional Reach Distance Requirements")

Figure 5.9 Maintenance access.

give detailed definitions of grasp reach limits in three dimensions (the seat mount controls illustrated in Figure 5.11 demonstrate one method of achieving this).

2. *Temporal sequence of use*: If controls are typically operated in a specific sequence, then they should be arranged in that sequence.
3. *Functional grouping*: Controls and displays which are related to a particular function should be grouped together. Figure 5.12 provides an example in which controls related to vehicle movement are grouped on the left, and those related to tool function grouped on the right.
4. *Frequency of use*: The most frequently operated controls should be located within close reach of all users.
5. *Importance*: Critical controls (even if infrequently operated) should be placed within easy reach of all users (e.g., emergency stop).
6. *Consistency*: The layout of controls in similar classes of equipment should be consistent. Some standards exist: for example, MDG 1, 4.3.6.1 specifies, "Machine motion controls should generally be left hand operated and working tool controls should generally be right hand operated" (New South Wales Department of Primary Industries, 1995).
7. *Spacing of controls*: Optimal spacing will minimise limb movement travel time whilst minimising the risk of inadvertent operation. Some guidance is available (e.g., see US Department of Defense,

Figure 5.10 Load-haul-dump (LHD) maintenance access.

1999, clause 5.4.1.3.7); additional protection against inadvertent control may also be appropriate through recessing controls, locking, guarding, or requiring two-handed operation for high-risk functions.

8. *Location compatibility between controls and associated displays*: Performance is improved if there is a logical arrangement of the location of controls and the displays to which they correspond. This may be achieved through proximity, or through ensuring that the layout of controls is consistent with the layout of the associated displays. Labeling should be considered a secondary cue.
9. *Optimal display location*: The most frequently observed displays should be located directly in front of the operator. The optimal height of displays is approximately 15° below horizontal eye height (Burgess-Limerick et al., 2000). Locating displays above horizontal eye height

Figure 5.11 Dozer rotary seat and controls.

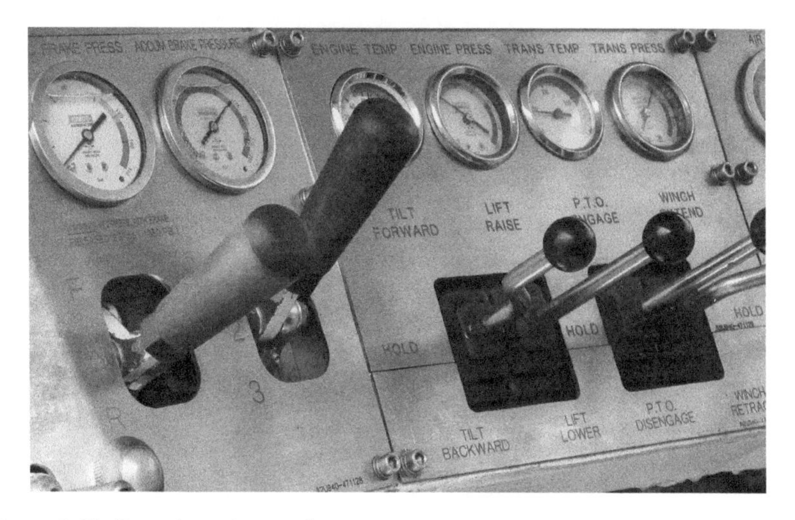

Figure 5.12 Grouping of controls.

should be avoided if at all possible (Figure 5.13 and Figure 5.14 provide examples of designs in which visual displays are located higher than is desirable; viewing the video camera displays shown in Figure 5.13, or the gauges in Figure 5.14, will require the operator to adopt an uncomfortable head and neck posture). Glare sources should be considered in the placement of visual displays, and backlit displays are desirable.

5.3.4 Visibility

As is also found in other domains (e.g., maritime or road transport), the information utilised during the operation and maintenance of mining equipment is predominantly visual. The size of mining equipment and design constraints imposed by the environment frequently result in restricted visibility for operator and maintainers (Figure 5.15). The consequences of restricted visibility include the necessity to adopt awkward postures (e.g., Figure 5.16), and the increased likelihood of collisions between moving equipment and stationary objects or surfaces, other vehicles, or persons. Workstation design should aim to remove impediments

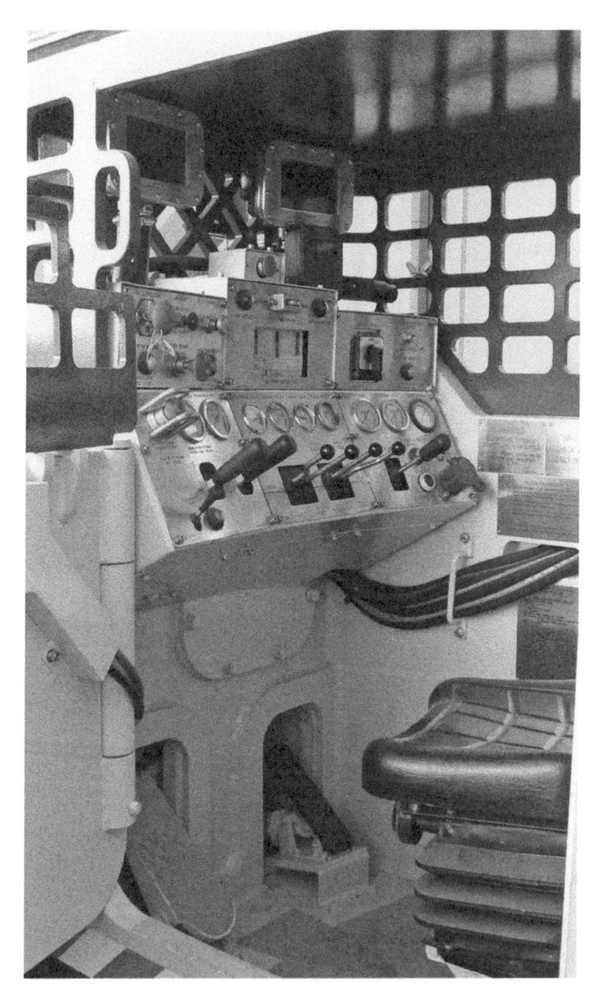

Figure 5.13 Video camera displays too high for comfortable viewing.

Figure 5.14 Gauges too high for comfortable viewing.

to visibility, and the evaluation of visibility should be a key consideration in design evaluation. This issue is of sufficient importance to justify more detailed treatment in a separate chapter later in this book (Chapter 7). Other control measures to compensate for the restricted visibility associated with many pieces of mining equipment include provision of video cameras, and proximity detection systems. These, and other new technologies, are addressed further in Chapter 9.

Figure 5.15 Chock carrier restricted visibility.

Figure 5.16 Chock carrier awkward postures.

5.3.5 Seating

Many workstations are designed to allow operation in a seated posture. Key considerations for the design of seating include range of height and fore–aft adjustability (and ease of adjustability) to ensure that optimal eye height and head clearance are obtained by all operators whilst maintaining appropriate access to pedals and hand controls; and provision of lumbar support, lateral support, and, for underground applications in particular, self-rescuer and cap-lamp battery. Where the sitting position is perpendicular to the direction of travel of the vehicle, a degree of seat rotation may be beneficial in reducing the awkwardness of neck postures required to operate the equipment. In other cases, there may be benefit in providing a seat which rotates 180°, such as the dozer seat illustrated in Figure 5.11, which allows the operator to always face the direction of travel. The vibration attenuation characteristics of seating provided on mobile plant and equipment (and consequently the requirement for the seat to match the weight of operators) also play a crucial role in reducing operators' exposure to whole-body vibration. This aspect will be considered in more detail later in Chapter 6.

5.4 Digital tools for workstation design

As the preceding discussion indicates, the task of accommodating human variability and human factors design principles is not an easy one. A range of digital tools exist to assist designers in meeting the challenges of designing workstations to suit the capabilities and limitations of the people who will operate and maintain the equipment, and to accommodate the anthropometric variability within these people. As well as visualising postural consequences of design decisions, such tools can also assist with assessment and documentation of other design aspects such as visibility. Chaffin (2008) provides a review of the history of these methods, their limitations, and future trends. In addition to the PeopleSize anthropometrics software already mentioned, other examples include SAMMIE CAD (2004), JACK (Siemens, 2010), HumanCAD (NexGen Ergonomics, 2009), and RAMSIS (Human Solutions, n.d.).

5.5 Conclusion

This chapter has considered issues associated with the layout of workstations and access to and egress from workstations, in particular mobile plant cabs, where space constraints frequently make accommodating anthropometric variability challenging. Principles for the layout of controls within workstations were addressed, whilst the design of control and displays themselves is addressed in Chapter 8. Seating and visibility were also noted as common difficulties in equipment design, and these issues are again addressed in Chapters 7 & 8.

chapter six

Physical environment and climate

Section 6.1 co-written with Robert Randolph
NIOSH, USA

Section 6.2 co-written with James Rider
NIOSH, USA

Section 6.3 co-written with Janet Torma-Krajewski
Colorado School of Mines, USA

Section 6.4 co-written with Tammy Eger
Laurentian University, Canada

As mentioned earlier, the mining environment is a challenging one where workers are exposed to many factors characteristic of a dynamic and hazardous workplace. In such a dynamic work environment, the physical attributes of the workplace change throughout the workday. In the case of mining, these workplace changes coincide with the production of materials, that is, the environment changes as more materials are removed and a new environment is revealed. This environment can be unpredictable, uncontrollable, and unexpected (Steiner, 1999; Sharf et al., 2001). Designing equipment and workplaces in this environment can therefore be difficult. This chapter provides basic concepts to detect and design for the deficiencies in the mining environment involving noise and hearing, dust, heat and cold stress, and whole-body vibration. Chapter 7 then continues this by introducing basic concepts in vision and lighting to design for visual deficiencies in mining. For both of these chapters, each factor is described in terms of how the worker is affected by it, what aspects of mining present these factors, how to measure it, and how to reduce any potential impact of these factors.

6.1 Sound and hearing

Noise is a significant hazard in most sectors of the mining industry with its most evident effect being noise-induced hearing loss (NIHL). In 1971, the US Bureau of Mines measured hazardous noise and found that 73 percent of underground coal miners were exposed, and later found that miners had measurably worse hearing compared to the national average for all occupations. More recently, Tak et al. (2009) reported mining to have the highest incidence of hazardous noise exposure (76 percent) of all industries. Hearing loss is the obvious outcome, but exposed workers are more likely to experience speech interference, disturbed sleep interference, excess stress, tinnitus, and decreased work performance (Kryter, 1994). By age sixty-five, over half of the mining workforce has hearing loss to the point of causing speech difficulty (National Institute for Occupational Safety and Health [NIOSH], 2009). This hearing loss can cause communication and safety concerns at work and at home.

Noise is often defined as unwanted sound. Noise-induced hearing loss can be temporary (up to sixteen hours, recovering within minutes or hours) or permanent, called respectively *temporary threshold shift* (TTS) or *permanent threshold shift* (PTS). Factors that influence both include exposure duration, exposure level, and individual susceptibility. One problem is that permanent changes can take years to accumulate before its effects become apparent, which can present difficulties identifying and avoiding the specific exposures that caused the hearing loss.

Noise-induced hearing loss usually starts in the frequency region around 4,000 to 6,000 Hz, the upper levels of the speech region. The first noticeable symptoms include difficulty hearing higher pitched sounds or voices such as female or children's voices and difficulty understanding certain consonant sounds.

It is possible to get an idea of how noise exposure can affect hearing over time. The NIOSH Hearing Loss Simulator is a Windows-based program that displays a control panel on the computer for playing sounds whilst adjusting simulated effects of noise. The individual's age can be programmed along with years of noise exposure, and the effects are shown on the frequency band control panel and the sound-level display screen whilst the user listens to the audio playback. This tool can be downloaded from www.cdc.gov/niosh/mining/products/product47.htm

6.1.1 What is sound?

Sound is an auditory sensation that is due to pressure fluctuations of acoustic waves (oscillations about the ambient pressure of the atmosphere, consisting of longitudinal vibration of air molecules). For humans it is a variation of pressure in the ear, which can be detected by receptor cells in the ear. The two most important characteristics for us are frequency and level (which correspond in layman's terms to pitch and loudness).

In terms of frequency, humans are capable of detecting sound between 20 and 20,000 Hz, but we are most sensitive to sound in the range of 2,000–5,000 Hz (roughly corresponding to human speech).

Sound level is defined in terms of sound pressure level (SPL) and is measured in a logarithmic unit of decibels (dB). As dB is logarithmic, 100dB is not twice the level of 50dB; instead, a tenfold increase in level occurs with each 10dB increase.

The ear is more sensitive to some frequencies than others. For this reason weighting networks are often included in sound-level metres to approximate the way the human ear responds. Of the three various weighting networks, A-weighting is the most commonly used in hearing loss prevention practice because it can be used when estimating the risk of hearing damage.

6.1.2 *Hearing and age*

Most age-related hearing loss is caused by gradual deterioration of the (cilia) hair cells in the cochlea. The symptoms include difficulty understanding conversations, especially when background noise is present and hearing higher frequency sounds such as a female's voice. Normal hearing loss begins to occur at the rate of 1 dB/year for each year after 60. The combination of high noise exposures in the mining environment and the aging worker may speed up the onset of hearing loss. Workplace changes and equipment design that reduce noise exposure can help slow down this process. Other design factors can help accommodate workers who have lost hearing due to aging and noise exposure and ensure that safety messages are communicated. For instance, to ensure that people can hear sounds and alarms, provide redundant signals such as alarms and flashing lights to alert people about equipment location. These redundant signals will also help workers with normal hearing when auditory signals are masked by ambient noise.

6.1.3 *Variables of noise exposure*

The combination of noise level and duration that create exposure can be measured using a noise dosimeter. This device can be worn by a worker for a full shift and then the total noise exposure is computed at the end.

MSHA's 30 CFR Part 62 is the noise rule which established noise exposure limits for mine workers in the USA. According to the noise rule, an individual's occupational noise exposure should not exceed an 8-hour time-weighted average (TWA) of 90dB(A), which is defined as the permissible exposure level (PEL). Higher levels are permitted but for shorter time periods. The TWA action limit (AL) is 85dB (A) where hearing protection devices must be provided to workers. A TWA of 105 dB (A) or more requires dual hearing protection such as a muff over an earplug. A TWA of 115 dB (A) is not permitted. Similar requirements are contained in standards such as AS1269:2005.

6.1.4 Noise protection strategies

High noise levels are inevitably associated with the use of mining equipment, however, there are a variety of control strategies that can be employed. There are three considerations involved in developing effective noise controls: source, path, and receiver. These three elements react together to form a unique situation; the sound generated by the source can be controlled or reduced and the path the sound travels can be interrupted to reduce the exposure reaching the receiver. For example, noise is commonly caused by vibrating machinery, and isolating the vibrating machine from surrounding structures can reduce the noise transmitted through the structural path. If a noise cannot be sufficiently reduced through noise controls, earplugs and/or ear muffs may be necessary. These PPE vary greatly and need to be correctly chosen for the environment and be worn properly. These devices also reduce the ability to communicate and hear sounds necessary for performance and safety. From most effective (but generally more costly engineering solutions) to least effective (but often cheapest) they are:

1. Engineering controls that reduce noise at the source or prevent noise from reaching the worker's location;
2. Administrative controls that reduce the worker's exposure through changes in procedures or behaviour to reduce their time and proximity to noise sources;
3. Hearing protection devices that reduce the noise reaching the worker's ears.

Table 6.1 shows the EMESRT Noise Design Philosophy for surface mining equipment. Of key interest here are the examples of industry attempts to mitigate risks (for example, the noise absorbent material in seals of large mobile equipment shown in Figure 6.1 below).

6.1.5 Noise: Summary

To summarise, the effects of noise in mining can be threefold:

- At extreme levels it can be dangerous, causing deafness.
- At moderate intensities it is more likely to affect performance—often due to interference with hearing. Some warnings may not be

Table 6.1 Earth Moving Equipment Safety Round Table (EMESRT) Noise Design Philosophy for Surface-Mining Equipment

Noise	
Objective	The objective is to minimise risks related to noise generated by the equipment to ALARP, including consideration in design for foreseeable human error.
General outcome	The intended design outcome should include the following: Engineered controls which, during normal operation of the plant and with the air conditioner or heater on the "high" setting, will maintain noise levels within the operator environment below occupational exposure limits (OELs) as defined in AS 1269 and ISO 6395 (International Organization for Standardization [ISO], 2008) 80dbA—tested by AS 1269, with a preference for machines that do not exceed 75dB(A) LAeq, taking into consideration normal operating conditions such as under full power and audible radio
Risks to be mitigated	1. Risk of noise-induced hearing loss to operators 2. Risk of noise-inducing hearing loss in workshop environs 3. Risk of distraction to the operator from excessive noise 4. Risk of distraction to the bystander through noise in the environment 5. Risk of excessive noise impacting the ability of the operator or bystander to hear warnings or alarms 6. Risk of noise-induced fatigue
Examples of industry attempts to mitigate risks	a. Enclosed and tightly sealed and pressurised air-conditioned cabins b. Thicker sound material, stronger access panels, and additional insulation to the cab c. One-piece dual-pane glass (e.g., toughened, laminated, and shatterproof) on all sides that significantly reduces the operator's sound exposure. d. Door seals that are positioned so that they are not prone to physical damage in normal operation e. Selection and relocation of air-conditioning systems to reduce noise

(*Continued*)

Table 6.1 Earth Moving Equipment Safety Round Table(EMESRT) Noise Design Philosophy for Surface-Mining Equipment (*Continued*)

	Noise
	f. Sound suppression and absorption materials around outside components (exhaust system, engine compartments, and cooling fans)
	g. Active noise-cancelling devices designed to lower noise caused by low-frequency sound waves
	h. In-cab communication headsets with active listening technology designed to integrate all radio communication (company and AM/FM) directly into the headset and limit noise output

Figure 6.1 Noise-absorbent material in seals of large mobile equipment.

recognised or heard which can result in performance decrements or accidents and injuries.

- At low levels, noise can reduce comfort and increase annoyance (so, potentially, influencing work performance and well-being through, for example, lowering of a maintenance operator's concentration).

6.2 *Dust*

6.2.1 *Breathing and dust*

Fresh air enters the lungs, and eventually into tiny air sacs called *alveoli*, where oxygen from the inhaled air is transferred to the blood, and carbon dioxide in the blood is transferred to the air. However, as air enters the upper and lower airways it passes over a sticky mucus layer that lines the cavity. When foreign particles such as dust or bacteria come in contact with the mucus, these particles get trapped and are removed from the inhaled air. High concentrations of respirable dust over an extended period of time can overwhelm and damage the lungs, causing serious health problems.

Dust in general is measured in micrometres, commonly known as *microns*. From a size perspective the width of a single strand of a human hair is 50–75 microns and the human eye can see particles as small as 40 microns. Respirable dust is defined as any airborne material that measures less than 10 microns and penetrates the upper and lower airways into the alveoli region of the lungs.

Excessive or long-term exposure to harmful respirable dust may result in a respiratory disease called pneumoconiosis. Coal workers' pneumoconiosis (CWP), commonly referred to as *black lung*, is a chronic lung disease that results from the inhalation, deposition, and buildup of coal dust in the lungs, and the lung tissue's reaction to its presence. In addition to CWP, coal mine dust exposure increases a miner's risk of developing chronic bronchitis, chronic obstructive pulmonary disease, and pathologic emphysema.

With continued exposure to the dust, the lungs undergo structural changes that are eventually seen on a chest X-ray. In the simple stages of disease (simple CWP), there may be no symptoms. However, when symptoms do develop, they include cough (with or without mucus), wheezing, and shortness of breath (especially during exercise). Figure 6.2 shows

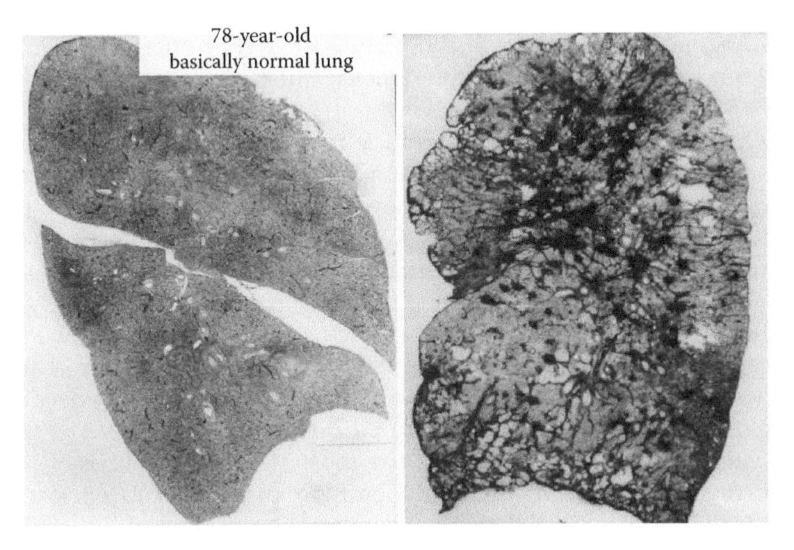

Figure 6.2 Normal lung (left) and a lung from a miner diagnosed with coal workers' pneumoconiosis (CWP; right).

a normal lung and a lung from a miner who has been diagnosed with CWP. In the more advanced stages of disease, the structural changes in the lung are called *fibrosis*. Progressive massive fibrosis (PMF) is the formation of tough, fibrous tissue deposits in the areas of the lung that have become irritated and inflamed. With PMF, the lungs become stiff and their ability to expand fully is reduced. This ultimately interferes with the lung's normal exchange of oxygen and carbon dioxide, and breathing becomes extremely difficult. The patient's lips and fingernails may have a bluish tinge and there may be fluid retention and signs of heart failure. If a person has inhaled too much coal dust, simple CWP can progress to PMF.

Silicosis is a form of pneumoconiosis caused by the dust of quartz and other silicates. The condition of the lungs is marked by scarring of the lung tissue (nodular fibrosis) which causes shortness of breath. Silicosis is an irreversible disease; advanced stages are progressive even if the individual is removed from the exposure.

6.2.2 Dust control in mining

In underground coal mines, airborne dust concentrations are typically the highest for workers involved in the extraction of coal at the mining face. Longwall shearer operators, jacksetters, continuous miner operators, and roof bolters are occupations with elevated potential for being exposed to excessive levels of respirable coal mine dust. Workers in some above-ground coal-mining operations also have increased exposure to coal mine dust. These include workers at preparation plants where crushing, sizing, washing, and blending of coal are performed and at tipples where coal is loaded into trucks, railroad cars, river barges, or ships.

Outby dust sources such as vehicle movement, removing stoppings, and delivering and unloading supplies can contribute significantly to worker dust exposure at the longwall face. Dust generated by these sources enters the ventilating airstream and remains airborne across the entire face, which can have a significant impact on the dust exposure of all face personnel. Efforts must be made to maximise the quantity and quality of ventilating air that reaches the face area. Operators have to be diligent in monitoring moisture content of the dust on intake roadways. Properly maintaining the belts is one of many vital components needed to keep respirable dust levels low along the belt entry.

The stageloader or crusher is the most significant source of respirable dust in the headgate area. Common practise is to enclose the stageloader and crusher through a combination of steel plates, strips of conveyor belting, and/or brattice. With the quantity of coal being transported through the stageloader or crusher, it is imperative that all seals and skirts be carefully maintained to ensure that dust stays confined within the enclosure. At a minimum, water sprays should be placed on both sides of the crusher and at the stageloader-to–section belt transfer (Rider and Colinet, 2007).

Water spray application is the primary control being used to substantially reduce dust liberation during longwall mining. Shearer cutting drums are equipped with drum-mounted water sprays which apply water directly at the point of coal fracture to maximise dust suppression and add moisture to the product to minimise dust liberation.

Respirable dust from a continuous miner affects the operator as well as anyone working downwind of the active mining area. In the United States, two types of ventilation systems are used to supply fresh air to the face of a continuous-mining section. Typically, the most effective method from a dust control standpoint is exhaust ventilation. Fresh air is brought to the face in the working entry, and line brattice or tubing is installed within the entry to create an air separation. Dust-laden air is then drawn from the face through the tubing or behind the curtain. The second system is blowing ventilation, where clean air is brought to the face behind the brattice or in tubing and discharged towards the face. Dust-laden air is then carried out of the face through the entry. This type of ventilation typically penetrates deeper towards the face and is more effective for methane control.

The largest source of roof bolter operators' dust exposure can occur when working downwind of the continuous miner. The mining pattern should be designed to eliminate or at least minimise the need for the bolter to work downwind of the continuous miner. Other sources of dust exposure for the bolter operators are the drilling process and maintaining the dust collector. The majority of the roof bolters in the United States utilise a dry vacuum dust collector system that pulls dust through the drill to a dust collector box. The dust is removed from the airstream and deposited in chambers of the collector box or captured by a filter cartridge. Properly maintaining the dust collector system is critical; checks should be made for loose hose connections and damaged compartment door gaskets. Vacuum pressure at the drill head should be checked periodically for proper airflow. The operator should maintain an upwind position when removing dust from the dust box to reduce exposure.

Workers are located in an operator's booth, control room, or enclosed cabs at many mineral-processing plants to give them a safe work area and to isolate them from dust sources. Enclosed cabs can also be used on surface mining equipment to reduce dust exposure. If these areas are properly designed, they can provide very good air quality to the worker. The most effective technique for reducing operator exposure is with filtration and pressurisation, which should have the heating and air-conditioning (HVAC) components tied in as an integral part of the system. The following key factors necessary for achieving an effective enclosure filtration and pressurisation system were identified:

1. Ensure good cab enclosure integrity to achieve positive pressurisation against wind penetration into the enclosure (Cecala et al., 2005).
2. Utilise high-efficiency respirable dust filters on the intake air supply into the cab (Organiscak and Cecala, 2008).
3. Use an efficient respirable dust recirculation filter (Organiscak and Cecala, 2008).
4. Minimise interior dust sources in the cab by using good housekeeping practices, such as periodically cleaning soiled cab floors, utilising sweeping compound on the floor, or vacuuming dust from a cloth seat. Also, relocate heaters that are mounted near the floor.
5. Keep doors closed during equipment operation. Studies showed a ninefold increase in dust concentrations inside the cab when doors were frequently opened during the sampling period (Cecala et al., 2007).

6.2.3 *Respiratory protection and other personal protective equipment*

Engineering controls should be used where feasible to reduce workplace concentrations of hazardous materials to the prescribed exposure limit. However, some situations may require the use of respirators to control exposure. Respirators must be worn if the ambient concentration of coal dust exceeds prescribed exposure limits. Respirators may be used

- before engineering controls have been installed,
- during work operations such as maintenance or repair activities that involve unknown exposures,
- during operations that require entry into tanks or closed vessels, and
- during emergencies.

Workers should use appropriate personal protective clothing and equipment that must be carefully selected, used, and maintained to be effective in preventing skin contact with coal dust. The selection of the appropriate personal protective equipment (PPE) (e.g., gloves, sleeves, and encapsulating suits) should be based on the extent of the worker's potential exposure to coal dust. To evaluate the use of PPE materials with coal dust, users should consult the best available performance data and manufacturers' recommendations.

To summarise, the effects of dust in mining are as follows:

- Coal mine dust exposure increases a miner's risk of developing chronic bronchitis, chronic obstructive pulmonary disease, and pathologic emphysema.
- Silicosis, a form of pneumoconiosis caused by the dust of quartz and other silicates, is an irreversible disease; advanced stages are progressive even if the individual is removed from the exposure.
- Dust issues occur both underground and at the surface.

There are many interventions to reduce exposure to dust at the source and on the person.

6.3 *Heat, cold, and climate control*

Both surface and underground mines present environmental heat or cold stress loads that require interventions to maintain the health, safety, and performance of miners. In any thermodynamic system, a system that is concerned with the exchange of heat work and matter in an environment, there exists a range of temperatures that results in optimal performance. The human body, as a thermodynamic system, has a very small range of temperatures (36–38°C) that allows for optimal performance. Slightly beyond this range of temperatures, the continued existence of the human body is challenged and death may occur. The primary focus of this section will be upon heat stress, given that it is usually more of an issue in mining than cold stress.

6.3.1 *Extent of the issue*

The gold mines in South Africa have experienced perhaps the greatest incidence of heat-related illnesses in the mining industry. As shown in Figure 6.3, prior to 1940 as high as twenty-six fatalities per year from heat stroke alone occurred in these mines. Fortunately, preventive measures were initiated and the number of fatalities per year dropped significantly to fewer than ten after 1962, with the exception of one year, 1983 (Kielblock and Schutte, 1993). Van der Walt (1981) also reported that between the years 1969 and 1980, 205 heat stroke cases were reported in South African gold mines, with thirty-nine of these cases resulting in fatalities.

Donoghue et al. (2000) reported on heat exhaustion cases occurring during a one-year period in five underground metal mines located in Australia; three of the mines operated at depths less than 1,200 metres, and two of the mines operated at depths greater than 1,200 metres. At these mines, the incidence rate was 43.0 cases/million man-hours. During the month of February (generally the hottest month in Australia), the rate

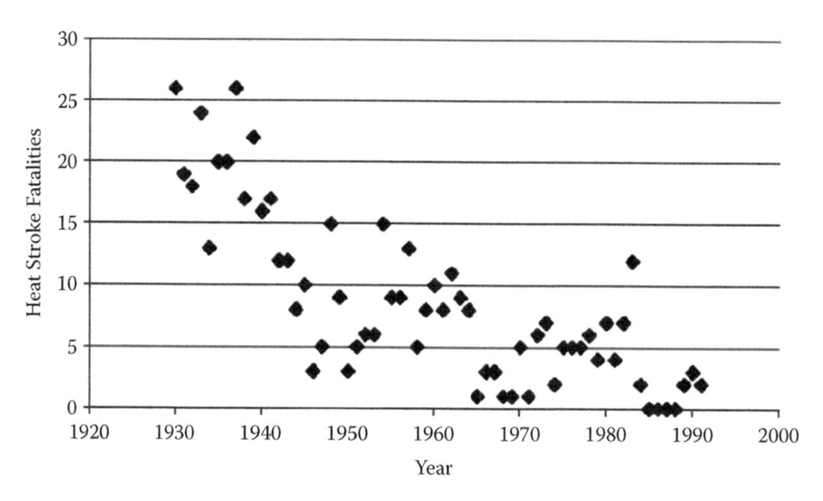

Figure 6.3 Number of heat stroke fatalities from 1930 to 1991 occurring in South African gold mines (adapted from Kielblock and Schutte, 1993).

was 147 cases/million man-hours. For the three mines operating above 1,200 metres depth, the incidence rate was 18.4 cases/million man-hours. Below 1,200 metres in depth, the incidence rate was 58.3 cases/million man-hours. The ratio of incidence rates for mines operating below 1,200 metres compared to those operating above 1,200 metres was 3.17. Fifty-seven percent of the cases occurred at or near the face, where the ventilation was poor and humidity high. Forty-two percent of the cases occurred on the first day of a four-day work schedule after having four days off.

Donoghue (2004) reported on heat illness to the Mine Safety and Health Administration by the US mining industry during a nineteen-year period (1983 to 2001). During this time, 538 non-fatal and zero fatal heat illness cases were reported. The highest incidence rates occurred in mills and preparation plants (nonmetal, metal, and stone) and in underground metal mines. Most of the cases in underground metal mines occurred at the working face. The lowest incidence rate occurred in underground coal mines. The higher rates in the underground metal mines as compared to coal mines were attributed to metal mines being deeper than coal mines, having lower air velocities at the working face, and having higher metabolic rates associated with work tasks.

6.3.2 *Overview of environmental heat stress*

For the human body to maintain a safe body temperature during exposure to environmental heat stress, a constant exchange of heat between the body and the environment must occur. This heat exchange is governed by the fundamental laws of thermodynamics of heat exchange between objects. The amount of heat that must be exchanged is a function of the total heat produced by the body as metabolic heat, and the heat gained from the environment. The rate of heat exchange with the environment is dependent on air temperature and humidity, skin temperature, air velocity, evaporation of sweat, radiant temperature, and clothing worn. The basic heat balance equation defining this relationship is as follows:

$$\Delta S = (M - W) + C + R - E$$

where

ΔS = change in body heat content
M = heat gained from the body metabolism
W = heat lost from work performed
C = convective heat exchange
R = radiation heat exchange
E = evaporative heat loss

The three major paths of heat exchange between the body and the environment are convection, radiation, and evaporation. A fourth path, conduction, plays only a minor role when a body part has direct contact with an object, such as a tool, equipment, floors, furniture, and so on, and is not usually considered when determining the change in body heat content. Another minor path of heat loss or gain is from air exchange during respirations. In more depth, the three major paths of heat exchange are as follows:

Convection: Convective heat exchange occurs between the skin and the air layer immediately surrounding the skin, and is a function of the difference between air temperature and the mean weighted skin temperature, and the rate of air movement over the skin. Assuming a mean weighted skin temperature of 35°C (95°F), there will be convective heat gain if the air temperature is greater than 35°C, and convective heat loss if the air temperature is less than 35°C.

Radiation: Radiative heat exchange occurs when a temperature gradient exists between the mean radiant temperature of the surroundings and the mean weighted skin temperature.

Evaporation: Evaporation of sweat from the skin is the main cooling mechanism for the body. The evaporation rate is a function of air movement over the skin, and the difference between the water vapor pressures of the air and the skin.

The presence of clothing impacts the heat exchange between the body and the environment because it acts as a barrier for convective, radiative, and evaporative heat exchange. The greater the thickness and impermeability to vapor and air of the clothing material, the greater the interference with heat exchange. When the clothing differs in the type, amount, and characteristics of typical work clothing (single layer), a corrective factor must be applied.

6.3.3 *Environmental heat stress in mining*

Environmental heat stress conditions found in the mining industry range from those that are seasonal conditions found at surface mines to those that are constant challenges that require extreme engineering controls along with rigorous administrative controls. In surface mines, radiant heat gain from the sun is the main contributor to heat stress conditions. Another source can be hot machinery, and humidity can add to the overall heat load.

In contrast, the heat load in underground mines comes from several sources. The primary contributors are the rock, the surface air, and autocompression. Heat from rock becomes a concern at depths greater than fifty feet (Misagi et al., 1976). Rock temperatures increase with depth as a function of the geothermal gradient, which depends on the specific location of the mine. At the Mount Isa Mines in Australia, the geothermal gradiant is 20°C/kilometre; the average rock temperature at the two-kilometre depth is 68°C (154.4°F) (Brake, 2002). Autocompression occurs when surface air is transported in vertical shafts to working levels; the air is compressed so that the volume is reduced but the amount of heat remains the same, resulting in air with a higher temperature. Autocompression will increase the air temperature 5.8°C/1,000 metres (5°F/1,000 feet) of shaft depth. Other sources of heat include equipment (diesel and electric), blasting, compressed-air service lines, air friction and shock loss, cement curing, oxidation of sulfides and timber, and broken rock. Groundwater and water used for dust control increase humidity levels, reducing the heat lost through evaporation.

Crooks et al. (1980) investigated heat stress levels in underground metal and nonmetal mines in the United States. The average dry bulb temperature was 16.6°C (62°F) and ranged from 2.8°C to 26.1°C (37°F to 79°F). Relative humidities averaged 70 percent, and ranged from 13 to 99 percent. Over 25 percent of the worksites included in the survey had relative humidities over 90 percent. Warmer, more humid, and slower airflows were found in production and development sites.

In South African gold mines, which are usually much deeper than the mines elsewhere in the world, over 50 percent of the total underground workforce worked in areas where the wet bulb temperatures were above 30°C (86°F). This value is equivalent to a dry bulb temperature of 31.1°C (88°F) and 90 percent relative humidity, or 33.3°C (92°F) and 80 percent relative humidity (Van der Walt, 1981).

6.3.4 Physiological responses to heat stress

When the human body is exposed to environmental heat stress, physiological responses occur that assist with maintaining the core body temperature between 36°C and 38°C. As the core temperature (temperature of internal organs measured by rectal temperature) rises, several organ systems including the central nervous system (CNS), the circulatory system, and sweating mechanisms respond to reduce the heat load.

The physiological responses to heat stress are enhanced or attenuated by numerous personal factors, including acclimation to heat stress (particularly by improved sweat rate responses), physical fitness, increasing age, gender, body fat, consumption of alcohol or prescription medicines, and the presence of certain diseases. Excessive heat loads can result in a number of health effects, some more serious than others. Conditions of less concern are heat cramps and heat exhaustion, whilst heat stroke can be a life-threatening situation requiring immediate medical attention.

In addition to the above primary disorders associated with exposures to heat stress conditions, there are secondary effects regarding a worker's ability to perform work tasks. Excessive heat stress reduces a worker's ability to carry out physically demanding tasks at levels normally performed without heat stress. Misagi (1977) demonstrated the relationship between ambient temperature and performance level of miners loading mine cars and drilling rock. As the ambient temperature increased, the performance level declined. Heat stress exposures also reduce the ability to remain alert during prolonged, monotonous tasks; decrease the ability to make immediate decisions; and contribute to frustration, anger, and other emotions. These effects can result in lower productivity levels and increased incidence of accidents. The incidence of unsafe behaviours was shown to be a function of wet bulb globe temperature (WBGT), with the relationship being U-shaped. The lowest incidence rates occurred between 17°C and 23°C (62.6°F and 73.4°F). Outside of this range resulted in an increasing incidence of unsafe behaviours (Ramsey et al., 1983).

6.3.5 Heat stress indices and thermal limits

Determining effective exposure limits for heat stress is challenging because of the many variables that impact heat exchange between the human body and the environment. Over the years, numerous indices have been developed that combined several factors into a single measure or assessment of the heat stress level. These indices generally include some or all of the following measures, with modifications for acclimation, clothing, and metabolic rate: dry bulb temperature, wet bulb temperature, globe temperature, and air velocity.

The 2009 Threshold Limit Value (TLV®) for heat stress, published by the American Conference of Governmental Industrial Hygienists (ACGIH), has a goal of maintaining the core body temperature within 1°C of the normal body temperature of 37°C (98.6°F). The TLV represents conditions to which workers may be exposed repeatedly without adverse health effects, as long as the workers are heat acclimatised, adequately hydrated, not medicated, healthy, and wearing a traditional work uniform (long-sleeved shirt and pants). The action level (AL) represents conditions for unacclimatised workers, and that merits the implementation of a heat stress management program. The TLV and AL are based on the WBGT, with adjustments for clothing and metabolic rate. Additionally, the TLV and AL assume similar WBGT values for work and rest break environments, and eight-hour workdays in a five-day workweek with typical breaks. Because of individual differences in responding to heat stress exposures, the ACGIH also recommends utilising measures of heat strain and provides guidance on specific heat strain limits, primarily related to heart rate, core temperature, sweating, weight loss, urinary sodium excretion, and symptoms of heat strain (ACGIH, 2009). The TLV has been criticized for being overly conservative, resulting in productivity losses; needing to estimate the metabolic rate, which is done with limited accuracy; and lacking sensitivity to wind speed over the skin (Brake and Bates, 2002b; Miller and Bates, 2007).

The Australian mining industry has been applying the Thermal Work Limit (TWL) Heat Stress Index to control heat stress exposures for several years, resulting in reductions in heat illness cases and lost productivity (Brake and Bates, 2000). The TWL is defined as the limiting sustainable metabolic rate with a body core temperature < 38.2°C (100.8°F) and a sweat rate < 1.2 kilogram/hour-1 (2.64 pounds/hour-1). It applies to workers who are healthy, adequately hydrated, acclimatised, and performing work that is self-paced. Self-paced workers are workers who regulate their own work rate, and are not subject to external pressures (peer or supervisor pressure or monetary incentives). The TWL provides a prediction of a safe maximum sustainable metabolic rate utilising measures of dry bulb, wet bulb, and globe temperatures; wind speed; and atmospheric pressure. Adjustments for clothing are also considered. Use of the TWL is limited to environmental conditions with calculated metabolic rates from 60 to 380 watts/metre-2. Estimations of metabolic rates are not required (Brake and Bates, 2002b). Recommended TWL limits and suggested interventions for self-paced work are provided in Table 6.2.

6.3.6 Controls: General

Exposures to environmental heat stress are generally controlled by utilising a combination of personal protective equipment, administrative controls, and engineering controls. Again, the hierarchy of control that

Table 6.2 Recommended TWL Limits and Interventions for Self-Paced Work

Interventions	TWL Limit (W/m^{-2})			
	> 220	140 to 220	115 to 140	< 115*
	Unrestricted	**Acclimatization**	**Buffer**	**Withdrawal**
No work restrictions.	X			
Ready access to water at all times (minimum 4 litres/worker).		X	X	X
Workers must be acclimatised.		X	X	X
Workers must not work alone.		X	X	X
Evaluation for dehydration at end of shift.			X	X
Minimum wind speed of 0.5m/sec^{-1}.			X	
If minimum ventilation is not met, reassign workers if possible.			X	
If work continues, a corrective action request must be authorised within 48 hours.			X	
Heat stress work permit (authorised by management) required prior to start of work.				X
Routine work not permitted.				X
Only work associated with safety emergency or environmental corrective actions permitted.				X

Source: Adapted from Brake and Bates (2002b).

*or DB > 44°C (111°F) or WB > 32°C (89.6°F); W/m^{-2}: Watts per metre squared.

was introduced in Chapter 2 is useful here: it is better to eliminate heat stress rather than rely on a lower level of control (e.g., administrative). The controls selected will, however, be dependent on the specific sources of heat, as well as the metabolic load required to perform tasks.

Engineering controls: The primary engineering control for decreasing convective heat gain and increasing evaporative heat loss is the use of air conditioning in work areas, control rooms, equipment cabs, and rest or break areas. If the air temperature is less than the skin temperature, then air movement can be increased with the use of fans or blowers. However, if the air temperature is greater than skin temperature, then increasing air movement could lead to convective heat gain. Reflective shielding is an effective control for radiative heat gain, demonstrating reductions by as much as 80 to 86 percent (NIOSH, 1986).

Personal cooling devices: Several different types of personal cooling devices are commercially available and include water-cooled garments, air-cooled garments, cooling vests, and wetted over-garments. Whilst these devices provide effective cooling capacity, there are limitations regarding weight, mobility, length of cooling period, comfort, and maintenance. Reductions in performance levels may also result from wearing such devices. Because of these limitations, personal cooling devices are generally used for tasks that do not involve heavy physical labour and high mobility requirements, such as operating mining equipment.

Administrative controls: For some heat stress exposures, the implementation of administrative controls is sufficient to prevent heat disorders; however, for other exposures, administrative controls are used in conjunction with engineering controls and personal protective equipment. Examples of lower level controls include encouraging water intake, adjusting work schedules, establishing an acclimatisation program, screening for heat intolerance, providing training, requiring a buddy system, and rotating jobs.

6.3.7 Controls: Specific to mining

Because mining often involves processes and environment unique to this industry, there are additional considerations for controlling heat stress exposures in mines.

Blasting: Both heat and water vapor are released during blasting. Effective blasting will reduce undesirable release of heat into the mine environment, and blasting during off-shifts will reduce the heat load to miners (Misagi, 1976).

Deep underground mining: Because environmental conditions may exist in deep underground that are inhabitable by humans, cooling efforts may require multiple approaches. For example, the Enterprise Mine at Mount Isa (mining depth is 2,000 metres) utilises surface bulk air cooling, underground air cooling, and chilled service water (Leveritt, 1998).

Oxidation of minerals and/or timbers: Oxidation can only occur if oxygen and fuel are present. Sealing off worked-out areas where sulfides and timber are present can reduce oxygen levels and the potential for oxidation to occur (Misagi, 1976).

Groundwater: Groundwater, which is generally the same temperature of the rock, adds to the humidity and decreases the evaporative heat loss. The effect of groundwater on environmental heat stress levels can be controlled with transporting the water in covered channels or insulated piping to the surface (Misagi, 1976).

Curing cement: Curing of cement that is used for filling or shotcreting (rock bolt support) is an exothermic reaction, and can add to the heat load whenever this process occurs. Generally, the release of heat occurs over a few days and then the cement attains the temperature of the rock, but if the use of cement is an ongoing process, then the added heat load becomes a constant source that needs to be addressed usually with increased ventilation (Misagi, 1976).

Water and compressed-air lines: Air gains heat when it is compressed, which is then transferred to mine air. Running compressed air lines in the exhaust shaft or cooling the compressed air before it enters the mine reduces the heat load reaching the working areas. Hot water lines should be insulated and also routed through the exhaust shaft (Misagi, 1976).

6.3.8 Cold stress

In comparison to environmental heat stress, the human body has little ability to adapt to cold exposures. The primary responses to cold include vasoconstriction of peripheral blood vessels, and increases in metabolic rate and shivering, which start at a core temperature of 36°C (96.8°F). Workers should be protected from cold exposures that result in core body temperatures below this temperature. Physiological responses at lower core body temperatures include reduced mental awareness, reduction in decision-making ability, poor muscle coordination, or loss of consciousness, which could lead to fatal consequences. Hypothermia develops at a core temperature of 35°C (95°F) and then progresses toward a life-threatening state as the core temperature decreases further. Other less serious cold injuries that are more common include frostnip (loss of blood flow to the fingers, tips of the ears, and end of the nose with the skin turning a pasty, white color and losing sensation), superficial frostbite (formation of ice crystals within skin cells) and severe frostbite (formation of ice crystals in deeper layers of muscle and bone resulting in permanent tissue damage, often resulting in amputation of affected body parts).

Controlling exposures to cold often utilizes a combination of engineering, administrative, and personal protective equipment. The objectives of the Threshold Limit Value for cold stress is to prevent the core temperature from dropping below 36°C (96.8°F) and to prevent cold injuries to the extremities (ACGIH, 2009). Specific recommendations of this TLV® include:

- For tasks that involve handwork and air temperatures less than 16°C (60.8°F), the hands should be warmed (radiant heaters, contact warm plates, gloves).
- Tool handles should be covered with thermal insulating material when temperatures fall below -1°C (30.2°F).
- Insulating dry clothing should be worn in air temperatures below 4°C (40°F).
- For environments with wind, drafts, or artificial ventilating equipment, work areas should be shielded from the air movements or workers should wear windbreak garments. The equivalent chill temperature relating dry bulb air temperature and air velocity should be used to determine appropriate controls.

6.3.9 *Summary*

To ensure the safety and health of workers who are exposed to environmental heat or cold stress in mining, whether it is seasonal or a daily exposure, it is recommended that a comprehensive approach be followed. Such an approach considers all three types of controls discussed above. Implementation of those that are effective will reduce heat load during heat stress and maintain body temperature during cold stress, and/or prevent the occurrence of heat and cold stress–related disorders. Additionally, workers should be trained regarding the effects of heat and cold stress exposures, how their bodies react to such exposures, measures they can take to enhance their tolerance levels where applicable, and how to work safely during such exposures. Finally, a medical surveillance program should be available to workers to assist with preventing and identifying heat-related disorders and medical conditions that preclude working in cold conditions.

6.4 *Vibration*

Oscillatory motion (vibration) of varying frequencies and amplitude is associated with any equipment with moving parts. If a person sits, or stands, on a piece of moving equipment, or holds an activated power tool, vibration is transmitted to the person's body. For example, seated mobile equipment operators are simultaneously exposed to vibration at the feet from contact with the floor surface, at the buttock from contact with the seatpan, at the back from contact with the seat backrest, and at the hands from contact with controls of the vehicle.

The consequences of this vibration exposure may vary from benign to potentially hazardous, depending on the frequency and amplitude characteristics of the vibration, the duration of exposure, and the body part concerned. Exposure to vibration can have short-term consequences such as motion sickness, and decrements in the performance of visuo-motor tasks; as well as long-term cumulative health effects including irreversible damage to nerves and blood vessels of the hand and arm; and damage to the spine resulting in back pain. The design, operation, and maintenance of mining equipment influence the likelihood and severity of the unwanted consequences associated with exposure of humans to vibration. This section will review basic information about vibration measurement and human response to vibration, and discuss vibration control options.

6.4.1 *What is vibration?*

Vibration is oscillatory motion of an object about a fixed point, and is transmitted through mechanical structures. The rate of change of displacement of the object in motion (its velocity) is constantly changing direction, and consequently the rate of change of the velocity (acceleration) is also constantly changing direction. Whilst vibration could be described in terms of any of these (displacement, velocity, or acceleration), it is usually measured by an instrument sensitive to acceleration (an accelerometre) and described in terms of acceleration (units – m/s^2).

The amplitude of vibration may be described in terms of the maximum acceleration (peak acceleration), or in terms of the root-mean-square (rms) acceleration (analogous to an average acceleration). For perfectly sinusoidal motion, the rms value is equal to the peak value divided by the square root of 2. The vibration dose value (VDV) is a more complex measure of vibration (involving the fourth power of acceleration) which gives greater weight to high peak values of acceleration, for example jolts and jars (units = $m/s^{1.75}$).

The effects of vibration on the human body also depend on the direction in which the vibration occurs. Whilst vibration typically occurs simultaneously in all directions, the measurement is typically made in three orthogonal directions. For the standing or seated person, motion in the fore–aft direction is defined as movement in the x direction (the x-axis), lateral movement is defined as the y-axis, and vertical movement as the z-axis (Figure 6.4).

Figure 6.4 Principal axes of vibration exposure illustrated for a load-haul-dump machine operator. Vertical vibrations are measured along the z-axis, fore–aft vibrations are measured along the x-axis, and side–side (lateral) vibrations are measured along the y-axis.

Regardless of the measure employed for describing the amplitude of the vibration, the frequency of the oscillations is expressed in cycles per second (Hz). All objects, including body parts, have a natural or resonant frequency. If an object is exposed to vibration at that natural frequency the object resonates, and the amplitude of the vibration is increased, with a corresponding increased probability of adverse consequences. If the body is exposed to vibration outside its resonance frequency, the vibration energy will be dampened by the structure, and the vibration will be attenuated. Therefore, it is possible that a vibration input can be amplified by one part of the body and attenuated in another. For example, vibration frequencies between 0.5 Hz and 80 Hz are important from a whole-body vibration perspective; however, frequencies up to 1000 Hz are important from a hand–arm vibration perspective. Standards such as ISO 2631-1-1997 (ISO, 1997) and AS 2670.1-2001 (Standards Australia, 2001) provide frequency weightings which vary according to the direction of the vibration, and specify that vibration amplitude measures should incorporate these weightings.

6.4.2 *Consequences of vibration*

6.4.2.1 *Motion sickness*

The adverse effects of exposure to low-frequency vibration (< 1 Hz) include motion sickness. Motions in the range of 0.125 to 0.25 Hz are most likely to result in sickness. It is believed that the symptoms (nausea, dizziness, etc.) arise as a consequence of conflict between motion signals arising from the eyes, vestibular system, and non-vestibular motion detectors (proprioceptors). Similar symptoms may arise in the absence of any real motion, such as the situation in which visual motion cues are provided by a vehicle simulator in the absence of corresponding motion. This phenomenon has implications for the use of vehicle simulators in training (see Chapter 11).

6.4.2.2 Visuo-motor performance

Exposure to vibration can lead to degraded performance of visuo-motor skills, such as might be required to operate mining equipment. In particular, vibration can interfere with the acquisition of information via the eyes, and the output of information via hand or foot movements. The effect is usually caused by movement of the body part affected (e.g., eye or hand). Redesigning the equipment to either reduce transmission of the vibration to the body part, or make the task less susceptible to the vibration by, for example, increasing the size of a display or reducing the sensitivity of a control, is indicated if these effects become apparent (Griffin, 2006). The optimal control sensitivity in conditions which involve the operator being exposed to vibration is likely to be lower than the optimal control sensitivity in static conditions (see Chapter 8). These effects are likely to be proportional to vibration amplitude and greatest for vibration frequencies of the order of 2 to 8 Hz.

6.4.2.3 Health effects: Peripheral vibration

People are generally exposed to vibration from a localised source or a source that affects the whole body, and the adverse effects on human health associated with exposure to vibration may be conveniently separated into effects associated with vibration of the periphery (hand, arms, and feet—hence peripheral, or hand–arm, vibration) and of the trunk (particularly the back) or whole-body vibration.

Miners who operate drills and bolting machines and workers who use impact tools and power–hand tools to maintain equipment are exposed to peripheral vibration. Prolonged and repeated exposure of hands to vibration (typically power tools) is associated with the development of a range of adverse vascular, neural, and connective tissue consequences and symptoms, which may be collectively termed *hand–arm vibration syndrome*. One known potential consequence of peripheral vibration exposure is vibration-induced white finger, which is characterised, as the name suggests, by blanching of the fingers, starting with the fingertips but potentially affecting all of one or more fingers. The symptoms are also associated with exposure to cold conditions, and are often accompanied by numbness and tingling. Reduced dexterity and strength have also been reported (Griffin, 1990). There is also evidence that workers who are exposed to vibration via the feet could also be at risk for similar health problems reported in the hands. *Vibration-induced white feet* has been used to describe tingling, numbness, and blanching documented in the feet (Hedlund, 1989; Schweigert, 2002).

Although it is not possible to define the limits of safe exposure, guidelines for the evaluation of hand-transmitted vibration exposure in terms of frequency-weighted acceleration and daily exposure time are provided in standards such as AS 2763-1988 (Standards Australia, 1988) and ISO 5349-2001 (ISO, 2001). Vibration at a wide range of frequencies is implicated (up to 1 KHz) with the maximum frequency weighting given to 10 Hz vibrations. Control measures for reducing the risk of adverse health consequences associated with peripheral vibration include the redesign of tools to reduce vibration amplitude and the limitation of duration of exposure to power tool use. This can be achieved in three ways: reduce the magnitude of the forces causing the power tool vibration, make the tool less sensitive to the vibrating forces, or isolate the vibration of the tool from the grip surfaces (Skogsberg, 2006). The use of gloves to attenuate vibration is unlikely to be an effective control because the attenuation provided by the gloves is least effective at the most important frequencies (Griffin, 1998).

6.4.2.4 *Health effects: Whole-body vibration*

A range of adverse consequences have been associated with prolonged exposure to whole-body vibration such as that occurring during the operation of mobile plant and equipment used in mining (Village et al., 1989; Kumar, 2004; Eger et al., 2006; McPhee et al., 2009). These include muscular fatigue, gastrointestinal tract problems, autonomic nervous system (ANS) dysfunction, impaired circulatory function, effects on female reproductive organs, headaches, and nausea. The most commonly reported, and perhaps the most important issue, however, is the potential for prolonged exposure to whole-body vibration to cause or facilitate degeneration of spinal structures, resulting in back pain. A range of epidemiological evidence is available to support the association of whole-body vibration with an increased risk of subsequent back pain (NIOSH, 1997; Bovenzi and Hulshof, 1998). There is also evidence to suggest that poor sitting posture increases the risk of back pain associated with whole-body vibration (Wikström, 1993) and mobile equipment operators are known to be exposed to both whole-body vibration and poor seated postures (Kittusamy and Buchholz, 2004; Eger et al., 2008).

The mechanisms through which this association occurs are not clearly understood. It is known (Pope et al., 1998) that the resonant frequency of the spine for vertical vibration in a seated position is in the range of 4–6 Hz. Bone responds to impact loading by depositing more bone, and the vibration may result in the thickening of the vertebral body endplates, resulting, in turn, in reduced blood supply to the avascular intervertebral discs. This reduction would reduce the possible rate of repair of these structures, and hence increase the probability of long-term intervertebral disc damage.

Again, whilst it is not possible to define the limits for safe exposure, standards such as AS2670-2001 (Standards Australia, 2001, Appendix B) and ISO 2631-1 (ISO, 1997) provide guidance in the form of a "caution zone" and a "likely health risk" zone for both weighted rms vibration and vibration dose values. In 2004, ISO (2004) introduced a new standard for the evaluation of human exposure to WBV, called ISO 2631-5, *Mechanical Vibration and Shock: Evaluation of Human Exposure to Whole-Body Vibration, Part 5: Method for Evaluation of Vibration Containing Multiple Shocks*. ISO 2631-5 was established to quantify health risks associated with WBV containing multiple shocks. More specifically, adverse health effects to the lumbar spine and the vertebral endplates were predicted based on seated WBV exposure. Another method for the evaluation of negative health effects associated with WBV exposure is outlined in the European Union Directive 2002/44/EC (European Union, 2002).

These standards may be used to assess the potential effects of whole-body vibration levels to which operators of mining equipment are exposed. For example, Figure 6.5 presents the results of an investigation of the vibration levels to which operators of 21 Load-Haul-Dump (LHD)

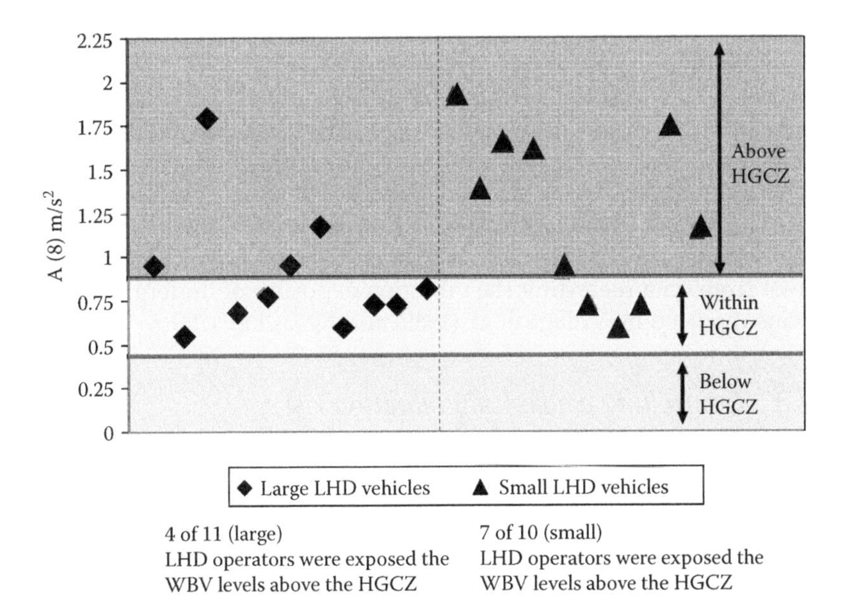

Figure 6.5 Smaller LHD vehicles (2.7 m³ bucket haulage capacity) exposed the operator to higher frequency-weighted rms acceleration levels leading to exposures above the ISO 2631-1 (ISO, 1997) health guidance caution zone for eight-hour exposure duration. Operators of 5.4 m³ bucket haulage capacity LHD vehicles were exposed to vibration levels within the health guidance caution zone.

Table 6.3 Operating Time Required to Reach Caution Zone for Selected Surface and Underground Mining Vehicles

Vehicle	Mean time to caution zone (hours)	Range (hours)
Surface dump truck	9	4–19
Surface loader	4	0.8–8.0
Surface dozer—ripping	2	0.6–4.0
Underground personnel carrier (passenger)	2	1–4
Underground equipment transport (no suspension)	0.4	0.11–2.00

Source: Data extracted from McPhee et al. (2009, Figure 11).

vehicles were exposed in terms of the Health Guidance Caution Zones (assuming an eight-hour exposure).

McPhee et al. (2009) presented the vibration characteristics of a wide range of vehicles in use at both surface and underground mines. The time taken for the use of the vehicle to reach "caution zone" and "likely health risk" levels was assessed according to AS 2670 (An extract is reproduced in Table 6.3) (Standards Australia, 2001). The findings are notable because, whilst illustrating the variability between equipment types and functions (such as, for example, surface personnel transport and dozer ripping) which might be expected, very large ranges were also found within similar equipment types and functions. This variability suggests that generalising the results from any particular vehicle to a class of vehicles would be problematic, and also highlights the potential contribution of variation in environmental conditions, and equipment design, use, and maintenance to the reduction of the amplitudes of whole-body vibration to which equipment operators and passengers are exposed. Similar conclusions may be drawn from data regarding the vibration exposure of underground LHD drivers reported by Village et al. (1989) and Eger et al. (2006).

6.4.3 *Controlling whole-body vibration risks associated with mining equipment*

Modifiable factors associated with whole-body vibration exposure related to mining equipment include the following: the environment in which equipment is operated, including roadway design and maintenance; activities performed using the equipment; equipment design, including suspension design and seat choice; equipment condition, including maintenance of suspension and seating; and operator behaviour, especially travelling speed.

The design and condition of roadways play a large role in determining the amplitude of vibration transmitted to the operator, and the importance of roadway maintenance cannot be over-estimated. Improving and maintaining roadway standards will have beneficial consequences in reducing damage to equipment as well as people. As well as scheduled roadway maintenance, it is important to ensure that any unexpected deterioration in roadway standards is promptly reported and corrected. This is particularly important in underground coal mines, where water issues can rapidly result in potholes—an extremely common cause of underground injuries (Burgess-Limerick and Steiner, 2006).

The activities performed have a large influence on vibration amplitude and direction. For example, ripping using a bulldozer leads to much higher vibration amplitudes than pushing. Village et al. (1989) reported higher vibration amplitudes for LHD operators during loading or dumping than during travelling. Vibration characteristics may also vary with the load carried by a vehicle. The implication is that vibration assessment of equipment must be undertaken for the range of tasks for which equipment is intended to be used, and the duration of exposure to tasks involving high amplitudes may require limitation.

The design of equipment, and particularly vehicle suspension, is a key area in which whole-body vibration exposure can be reduced at the source. Perversely, in the interests of ensuring equipment reliability, some mining equipment which operates in the roughest conditions has been provided with the least effective suspension (McPhee et al., 2009). For example, although underground shuttle cars were introduced in 1938, no cars were equipped with suspension until the last ten years, and shuttle cars without suspension may still be purchased.

The observations above (that vibration characteristics are specific to the tasks undertaken) and the roadway conditions encountered imply that the design of suspension should start with an assessment of these vibration characteristics under the whole range of operating tasks, conditions, and vehicle loads likely to be encountered. Design choices such as solid or foam-filled tyres also have implications for vibration characteristics.

Another aspect of the design of equipment is the choice of seating provided. As with design of vehicle suspension, ensuring an appropriate choice of seat requires a detailed understanding of the vibration characteristics which will be experienced at the seat during use. Providing an inappropriate seat can increase the vibration exposure of the operator. Boileau et al. (2006), for example, investigated vibration characteristics of sixteen underground LHDs. The dominant frequency components of the vibrations during use ranged from 2.6 to 3.4 Hz. A typical suspension seat provided for use on these vehicles was tested using a vibration simulator and found not to attenuate vibration at these frequencies, and indeed was more likely to amplify whole-body vibrations experienced under operating conditions.

Providing a seat which will attenuate vibration for the complete range of potential operator weights is also a challenge. The provision of weight adjustability is likely to be necessary; however, this usually implies providing a manual adjustment mechanism. To be effective, a manual adjustment mechanism must be easy to operate and remain so for the life of the seat under extreme environmental conditions. Even then, operator training in how and why to adjust the seat is necessary (operators frequently mistake the weight adjustment for height adjustment), and its effectiveness relies on the behaviour of operators. Automatic weight sensing and adjustment are more likely to be effective.

Regardless of their suitability when new, the vibration attenuation performance of both vehicle suspension and seating is likely to deteriorate with use. Appropriate maintenance and replacement schedules are required. A case can be made for the permanent incorporation of an accelerometre within the equipment seating and regular inspection of the data to identify when maintenance is required, before operators are exposed to potentially injurious vibration amplitudes.

Whilst a combination of the administrative and design controls discussed above has potential to reduce the exposure of equipment operators to whole-body vibration, the most effective control would be to separate the operator from the equipment via the use of remote control (either line of sight or non–line of sight) or automation. Remote control dozers are in use, as are automated LHDs in underground metalliferous mines. Of course, this is a larger issue with many human factors implications, and these will be dealt with in more detail in Chapter 9.

The next chapter will continue the approach used in this chapter and will examine issues specifically related to illumination and vision regarding mining equipment.

chapter seven

Vision, visibility, and lighting

Co-written with Tammy Eger
Laurentian University, Canada

The previous chapter dealt with a range of environmental issues (noise, dust, heat, and vibration) which have implications for the design, operation, and maintenance of mining equipment. Another critical environmental issue is the illumination of the mining environment.

Humans process visual data at about four times the rate of audible information. For mining, important audible data (i.e., verbal communications and audible warning alarms) are less useful given the noisy mine environment (Sammarco et al., 2009a). The visual ability to detect and recognise a hazard is affected greatly by the environment the target is presented. However, this sense of seeing can be enhanced by providing appropriate lighting and technology that increase one's visual performance. Illumination plays a critical role in the operation and maintenance of mining equipment because miners depend most heavily on visual cues to provide the information required to perform both operational and maintenance tasks related to equipment, and to detect potential hazards associated with, for example, falls of ground; slips, trips, and falls; and close proximity to moving machinery.

There are many human factors issues and illumination technology issues to consider. First, it is often difficult to detect hazards because of the low illumination levels inherent to mining (particularly underground), and the hazards are typically of low contrast and reflectivity. Glare is another factor to consider, as it can impede a miner's ability to identify hazards and perform a job safely. There are additional challenges for older workers because visual performance degrades as a person ages. Age is an important factor given an ageing mining workforce that averages about forty-four years. Poorer near vision as a result of reduced flexibility of the eye's lens begins by age forty (Eye Diseases Prevalence Research Group, 2004). Other contributing age factors include reduced pupil size, cloudier lens, and a reduction in the number of photoreceptors and depending on the visual task; visual performance declines 26 to 75 percent for older subjects. Lastly, illumination technology is changing with the greater use of light-emitting diodes (LEDs). The photometric and energy characteristics of LED sources differ in important ways. Traditionally, the lighting sources used in mining were incandescent, fluorescent, and some halogen dichroic, but now LEDs are becoming more popular. Accordingly, proper design of new mine illumination systems emerges as a critical factor to consider for miner safety.

7.1 Vision and lighting

Vision is one of the five basic senses (sight, smell, sound, touch, and taste) that people use to perceive their environment; it is generally considered to be the sense through which most of the information we use is acquired. Some environments cause us to rely more heavily on one sense than another as a consequence of the specific characteristics of the environment. For instance, a very noisy environment would force people to rely on vision more than hearing. Since the mining environment can be both noisy and dark, providing additional appropriate lighting is critical to enhancing vision. A brief understanding of how human vision works, its limitations, and its ability to adapt to different mining environments is presented below.

Understanding the functioning of the eye is the first step to adapting the environment and equipment to the worker's limitations and capabilities. Light enters the eye through the cornea and passes through the aqueous humour fluid. Then the light passes through the pupil, which is attached to the iris. The iris regulates the amount of light that enters the pupil by dilating (enlarging) or constricting it. Light rays pass through the pupil and are focused by the lens onto the retina (back inside surface of the eyeball). The retina gathers the light information and transmits it through the optic nerve to the brain for processing and perception (for example, determination of the form, size, shape, colour, position, and motion of the objects in view). (See Figure 7.1.)

Light detection is specifically performed by the rods and cones within the retina. These light-sensing receptors are responsible for visual resolution. Cones work better in high-lighting (photopic) situations and are used for daylight vision. They provide for colour vision. The rods dominate in low-lighting (mesopic and scotopic) situations, and they cannot detect fine difference in shape or detect colour. The rods are therefore more active in underground-mining (mesopic situations) and night-time surface operations. Since colour is difficult to discern in dark situations, the illumination must be increased to see colours or some other type of coding needs to be used to warn or alert workers to potentially hazardous situations. Reflective colours and materials are also sometimes used.

Adaption is the ability of the rods and cones to adapt to changes in illumination. When a person moves from a light area to a dark area (when

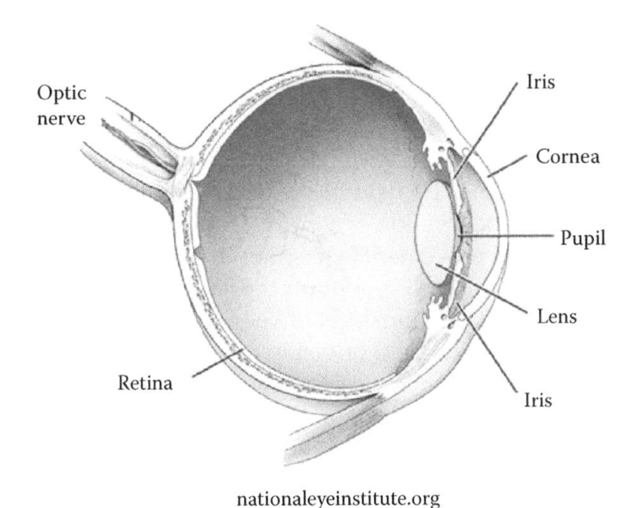

Figure 7.1 The eye.

going from the surface to underground or from daylight to darkness), it can take up to sixty minutes to fully adapt. When a person goes from dark to light the adaption is much quicker, typically one to two minutes. If an area is very bright in a dark environment, the environment will appear darker as the bright source dominates. This can happen when the bright light at a power centre is nearby or when a large piece of mobile equipment appears. This situation can make it difficult for workers to see hazards, such as tripping hazards or moving objects.

During transition from one area to another or from one type of walking surface to another, appropriate and additional lighting is needed to help recognise the change. Transitions are where many trip-and-fall incidents occur. These areas need to be particularly well lit to highlight the transition area and give more visual information to the worker.

Accommodation is the range at which individuals can focus within a given set of distances. The shortest distance the eye can bring something into focus is called the *short* (or *near*) *point*, whilst the longest distance is called the *far point*. For the ageing population, the near point begins to move farther away and the far point stays fairly constant. If proper illumination is provided, this far- and near-point range remains unchanged, but if lighting is poor, this range of accommodation becomes more narrow as the points begin to converge on each other and the focus is not as sharp. Contrast (described below) is the noticeable difference between a background and an object. More contrast allows for easier accommodation.

Field of vision is another component of vision which describes the area a person can see when both eyes and the head are stationary. If only the eyes rotate but the head remains stationary, the field of vision is less

(35° left and right of forward vision) than if the head can rotate whilst keeping the eyes fixed (60° left and right of forward vision). If no head or eye rotation can be accomplished, the normal line of sight can see 15° to the right or left with no head or eye movement. With both head and eye movement, this range increases to 95° on both the right and left (Sanders and Peay, 1988). Considering the field of vision is important when designing equipment because the location of controls and displays should be placed so that minimal head movement is needed. In addition, if an object is farther away it must be larger to maintain the same visual angle. Therefore, the visual angle is the relationship between the size and distance of the object as it determines the field of vision for the object. Visual angle and field of vision are both important when designing equipment and workplaces. For equipment to be designed properly, this component must be studied in the context of cab design and what is in the visual field. It is important that equipment structures are not blocking the operator's field of vision, including peripheral views. We will see that illumination plays an important role in enhancing the field of vision.

Peripheral vision is important for detection of movement. *Detection of movement* is the ability of a person to follow and learn information about an object by moving their eyes (tracking the object) or the ability to detect and learn information about an object which moves in front of them whilst their eyes are stationary. Both Martin and Graveling (1983) and Sammarco et al. (2009b) found that adding background light (with a spectral content having more of the short wavelengths of light) in addition to cap lamp lighting improved peripheral vision and the speed at which a person can detect a moving object.

7.2 Illumination and vision performance

As described above, the eye senses light to produce vision. Therefore, illumination is of paramount importance in mining when designing equipment and workplaces. The Illuminating Engineering Society of North America (IES) reports that the coal mine face is the most difficult lighting environment in the world (IES, 1993). The optimization of all aspects of lighting is nearly impossible, so many trade-offs must be made. For instance, underground mining poses an obvious problem for workers to have enough light to work but not too much to cause glare. Also, the battery-powered personal lighting provided to the miner has had limited range and contrast effectiveness as well as limited peripheral enhancement. For surface mines, sun glare can be a problem during the day, and during night operations, artificial lighting is often used similarly to underground mining.

The goal of lighting design is to allow people to be able to see what they need to see in an environment so that they can function and perform tasks safely and effectively. If the environmental lighting is improperly designed, there is a greater chance of error due to the inability to see and a greater chance of discomfort from eye strain and headaches. There is also a psychological component of being disoriented, confused, and even depressed. For mining tasks, lighting is critical to detecting hazardous situations. The lighting terminology and key concepts needed to understand illumination requirements are as follows:

Illuminance is the amount of light falling onto a surface. It is measured in lumens per square metre (cd·sr/m^2) or lux (lx). It can be physically measured by a device called a *light meter.*

Luminance is the amount of light emitted (reflected or leaving) from a surface. It is measured in candela per square metre (cd/m^2).

Reflectance is the ratio of the total amount of light reflected by a surface to the total amount of light incident on the surface. Not all surfaces reflect, but rather some absorb. Darker surfaces tend to absorb more light.

Glare is when there are very high levels of light or brightness in the visual field. Glare is a difficult issue in underground mining. There is a tendency to increase the lighting levels in a dark environment, especially on underground mining equipment. These increases can often cause glare, making it difficult to see objects in the area and difficult to adapt to changes in the light intensities. There are three main areas of glare: two are inherent to underground mining, and all three can be found in surface mining.

- *Blinding glare* describes effects such as that caused by staring into the sun. It is completely blinding and leaves temporary or permanent vision deficiencies.
- *Disability glare* describes effects such as being blinded by oncoming car lights, or light scattering in fog or in the eye, reducing contrast, as well as reflections from print and other dark areas that render them bright, with significant reduction in sight capabilities.
- *Discomfort glare* does not typically cause a dangerous situation in itself, though it is annoying and distracting at best. It can potentially cause fatigue if experienced over extended periods.

Contrast is the ratio of the difference of object luminance and background luminance to the background luminance. In order for an object to be seen, it must have contrast with its background. Increasing contrast can happen in several ways: by increasing the illumination, changing the reflectivity, or using directional lighting to make a shadow so the object's shape can be seen.

As mentioned earlier, it is often a balancing act to achieve adequate levels of lighting without causing glare. It has been shown that increasing light levels improves task performance, including the ability to see hazards in the workplace (Sammarco et al., 2009b). However, it is a trade-off between increasing illumination to improve contrast and performance against the increased likelihood of glare, which can be detrimental to performance. There are many accidents involving machinery; some of those could be attributed to improper lighting level, but since it is not reported as the direct cause of the accident it is often not primarily addressed. At times, lighting is suggested as part of a long list of causes and solutions to an accident but rarely is identified as a primary cause of an accident.

7.3 Standards for mine lighting

Some countries such as Australia have specific requirements regarding safe mine-lighting systems, especially for underground coal mines. These mines are required to provide their lighting in an "intrinsically safe" housing to prevent ignition and explosion. For a lamp to be explosion proof, any explosion triggered by the lamp's electrical activity is contained within the device. In addition, the device itself will not become hot enough to cause an explosion. There are also requirements for the amount of light in the mining environment, but these requirements vary widely from country to country so it is outside the scope of this book to review them all. However, there are international lighting guidelines for mining. The IES and the Commission Internationale de l'Eclairage (CIE) are two such organisations which emphasize the quality and quantity of light and provide formulas to determine whether glare has an effect on visual performance. The International Electrotechnical Commission (IEC) also has a standard for caplamps: IEC 62013–1, cap lights for use in mines susceptible to firedamp.

7.4 Recommended lighting levels

The recommended levels of lighting for any workplace are 200–800 lx for operating machinery or performing assembly or maintenance tasks. Increases in contrast are needed for ageing populations or where low reflectivity is the case (McPhee, 2005). Trotter (1982) researched illumination standards set by various countries for underground coal. In haulage ways, lighting levels between 0.5 and 21 lx were reported and between 20 and 80 lx around machinery. This is much lower than the recommended levels of 200–800 lx.

A generally accepted method for determining the attainment of lighting requirements in underground environments is by use of the Crewstation Analysis Program (CAP; Gallagher et al., 1996). This is a computer-based program that allows design and mining engineers to compare alternative lighting schemes for underground mines and mining equipment. The program is available through the Centers for Disease Control,

National Institute for Occupational Safety and Health (CDC NIOSH) mining program website (NIOSH, 2009).

7.5 Lighting used in underground mines

There are many traditional types of lighting used in underground mines, and all have their efficacy, efficiency, maintenance, and practical use issues. Below are some common sources of lighting and their efficacy range (see Table 7.1). At the time of this publication, it is estimated that the efficacy of the LED lighting has increased to 139 lm/W and will continue to increase over the next years as the technology is optimized. Research and development laboratory versions have reached 209 lm/W.

Given the limitations of illumination produced by a traditional miner's cap lamp, the miner's ability to identify potential hazards in a working area can often prove extremely difficult. According to the US Department of Labour, Mine Safety and Health Administration (MSHA), a total of 144 underground coal mine fatal accidents were attributed to moving mining machinery from 2001 to 2007. Twenty-five of those were attributed to accidents that involved continuous miners. The MSHA also reports that 11 percent of lost-time injuries reported from 2002 to 2006 were attributed to those caused by machinery. In addition to pinning and striking accidents, the second-highest cause of lost-time injuries is the slip or fall of a person. MSHA records indicate that slip or fall of person lost-time injuries account for 17.4 percent of the reported cases from 2002 to 2006 (Sammarco et al., 2009b).

Reduced levels of light around mining machinery can impair a miner's ability to judge the speed or direction of a machine. An inability to judge the machine movements and rate of travel can expose a miner to an extreme situation where the lighting may play a critical role in the ability to perform jobs safely. Increasing light levels around the perimeter of machines may potentially improve visual recognition of these hazards

Table 7.1 Luminous Efficacies for Various Light Sources as of October 2007

Light source	Typical luminous efficacy range (lm/W)
Incandescent (no ballast)	10–18
Halogen	15–20
Compact fluorescent (CFL) (includes ballast)	35–60
Linear fluorescent (includes ballast)	50–100
Metal halide (includes ballast)	50–90
Cool white LED 5000K (w/LED drive circuit)	47–64
Warm white LED 3300K (w/LED drive circuit)	25–44

Source: Adapted from the U.S. Department of Energy (2009).

and, consequently, reduce the number of injuries and fatalities in the mining industry. However, when increasing light levels, glare becomes a significant aspect to consider as it may negatively affect a miner's vision and ability to detect hazards. Accordingly, mine illumination systems must be properly designed to limit the effects of glare in order to effectively illuminate an area and improve safety. This is where recent developments in illumination—namely, LED lighting—can contribute to improved working conditions and potentially increase safety and reduce injury in mining.

As mentioned earlier, age affects visual acuity, and the average age of the mining workforce is increasing. The eye's ability to gather and focus light deteriorates in time. Therefore, it is imperative to develop a light source of sufficient quality to compensate for this loss of visual ability. A study by Sammarco et al. (2009b) compared the use of traditional lighting systems currently employed in the mining industry with the addition of LED lighting to those systems. The study concluded that significant improvements in detection of moving equipment and detection of tripping hazards were achieved by the addition of LED lighting to both incandescent and/or fluorescent lighting systems. This study engaged the use of auxiliary LED lighting by use of a prototype LED cap lamp and by addition of LED lighting to machine-mounted and surface-mounted applications. The results of the LED cap lamp study showed, for the older age group, a 15 and 23.7 percent improvement in moving and tripping hazards detection times, respectively. And, glare was measured as reduced by 45 percent. The study also concluded that when combining LED with fluorescent lighting in mounted applications, peripheral motion detection times improved by 12–14 percent. This improvement was attributed to LED's superior target illumination and contrast and its short wavelength content of visible light. An earlier spectral study showed that using light from the short wavelength end of the visible light spectrum leads to reduced disability glare.

Other important considerations to be factored into the development of an efficient lighting scheme for use in mining are location and direction of the light source, use of optics and shields to direct the light, and power consumption and durability of the light source. LED lights use approximately 75 percent less energy than incandescent light sources and provide a reported 25,000–1,000,000 hours of service life. Typically, the useful service life is defined by L_{30}, where the light has declined 30% from its initial value. Research is currently underway to critically compare LED lighting to other light sources. Research has also been done on the use of LEDs in the development of hazard warning systems. Therefore, when used in combination with other commonly used light sources, the addition of an LED light source can improve miner safety, lower overall maintenance costs, and lower energy consumption. Examples of these potential uses of LED lighting from current research studies are summarised below.

7.5.1 LED cap lamp

The results indicated significant improvements for older subjects when using a specific type of LED in the cap lamp (see Figure 7.2).

- Moving hazard detection improved 15.0 percent.
- Trip hazard detection improved 23.7 percent.
- Glare was reduced 45.0 percent.

These results provide important data for improving the design of future cap lamps and have the potential to positively affect miner safety.

7.5.2 Visual warning system (VWS)

NIOSH researchers have developed the machine-mounted VWS that warns people of impending machine movements and conveys the nature and direction of machine movement in order to reduce pinning and striking accidents.

Human subject test results indicate the following:

- 72 percent improvement in detecting machine movement with the visual warning system; the machine movement detection time decreased an average of 1.3 seconds.
- The system can be installed on any type of mobile mining equipment or can be used in other industries where moving equipment poses pinning and striking hazards.

Figure 7.2 LED cap lamp.

Figure 7.3 Wireless visual warning system (VWS).

7.5.3 Wireless visual warning system (VWS)

This wireless version is worn by people and warns them of impending machine movements (see Figure 7.3).

- Can be used with or without the VWS.
- Especially effective when workers are not looking at the machine.
- The system can be used for mining applications or by other industries where moving equipment poses pinning and striking hazards to workers.

7.5.4 LED area lighting

Human subject test results indicate an average 1.2-second improvement for detecting trip hazards in the area around a continuous-mining machine (see Figure 7.4).

7.6 Visibility and equipment design

Accidents involving mobile mining equipment can be reduced if equipment operators, pedestrians, health and safety officials, vehicle manufacturers, and management understand visibility limitations that accompany heavy mobile equipment operation. "Poor visibility" is a causal factor in serious incidents involving mobile equipment. For example, load–haul dump machines have struck pedestrians, other vehicles, equipment, and mine

Figure 7.4 Light-emitting diode (LED) area lighting.

walls and have driven into open "holes." In this vein, Godwin et al. (2008) found that a mobile equipment operator could not see a 5′6″ (168 cm) pedestrian twenty feet (6 metres) away from the machine cab for certain equipment designs; also, the mine floor became visible to the operator only ninety feet away from the cab.

7.6.1 Accident statistics

Canadian data published in 2001 by the Mines and Aggregates Safety and Health Association (MASHA) summarised accident statistics for load-haul-dump (LHD) vehicles in the Ontario mining industry over a fifteen-year period between 1985 and 2000. During this period, 1,690 accidents involving LHD vehicles were reported, and poor line of sight was identified as a causal factor in 24 percent of the accidents. Hitting other vehicles, hitting pedestrians, and hitting walls accounted for 16 percent of the accidents, and accidents due to unseen ground hazards (potholes, muck on the ground, or falling into holes) accounted for a further 8 percent of the accidents (MASHA, 2001). In the same time period, LHD vehicles were involved in accidents that resulted in ten fatal injuries to Ontario workers. The design of LHD vehicles, particularly ones with protective cabs, results in restricted sight lines and blind spots from the operator's position. Researchers have identified the bucket of mining vehicles as primary causes of blocked sight lines. Drivers and researchers have also noted that vehicle lights restrict sight lines and are often not placed to adequately light the roadway for operation (Boocock and Weyman, 1994; Rushworth, 1996). The machine characteristics responsible for blind spots, listed from most common to least common, were the cab and cab posts, lights and light brackets, boom, and air intake cylinder (Eger et al., 2004). One hundred and thirty LHD operators also provided feedback about the detection of objects in the environment (numbers in parentheses represent the percentage of LHD operators who found the object difficult to see): kneeling pedestrians (72 percent), equipment left on the ground (75 percent), standing pedestrians (39 percent), open holes (35 percent), parked vehicles (27 percent), and warning signs (15 percent) were difficult to see (Eger et al., 2004).

7.6.2 Strategies to improve line of sight from mobile equipment

Computer simulation software can be used to evaluate design changes to mobile equipment. For example, Godwin et al. (2008) evaluated line-of-sight improvements associated with several vehicle design changes. Improvements depended on how drastic the design change was, the size of the part being modified, and the size of the LHD vehicle the modification was evaluated on. On an individual-machine basis, large improvements of 15 percent or more were observed for wraparound window cut-outs, rounded and flattened engine covers, triangular bucket lips, and sinking air intake cylinders.

Furthermore, several machine characteristics have contributed to the restricted sight lines, including items such as cab posts, headlights, mudguards, engine covers, hydraulic units, and vent filters (Marx, 1987; Boocock and Weyman, 1994; Rushworth, 1996). Modifications to these components were able to enhance visibility as measured by Rushworth's Retrofit Index as well as reduce the individual operator risk scores by nearly 50 percent (Rushworth, 1996).

7.6.3 Cameras

Cameras can be used to aid in visibility by providing information that would otherwise be unavailable or missed. New technologies and applications of these technologies are being developed, and an introduction to these is provided later in this book. In Figure 7.5, a typical right-turn articulation for a load-haul-dump machine is shown. In the front right corner, a pedestrian and ground hazards are present; however, the pedestrian and hazard are not noticed in the eye view of the operator. In a recent study, Godwin and Eger (2009) showed an 80 percent improvement in operator line of sight when a combination of forward-facing and rear-facing cameras was installed on the vehicle.

7.6.4 Other visual aids

Whilst cameras are used in many situations, there are other visual aids to improve performance and safety. Secondary viewing devices such as closed-circuit television systems, radar or radiofrequency tags, ultrasonic sensors, laser detection, and Global Positioning System (GPS) technology have been

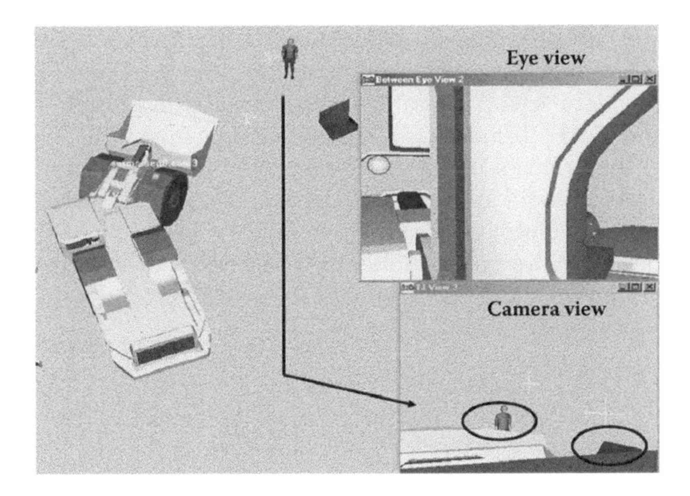

Figure 7.5 The field of view of the load–haul–dump machine operator is blocked by the cab post when making a right turn. A camera can allow the operator to see pedestrian and ground hazards.

implemented on industrial machines to aid in the detection of objects in blind spots (Ruff, 2001b). Each of these systems has challenges to overcome during implementation, especially in underground mining applications.

GPS technology cannot be used underground, whilst ultrasonic and laser sensors have not yet been proven as reliable for underground use (Ruff, 2001b). Radiofrequency identification systems with coloured light systems that illuminate based on the detected hazard have been implemented by some mining sites, but radars are highly susceptible to false alarms. False alarms or nuisance alarms are unacceptable and may lead to high levels of annoyance, reduced productivity, and a lack of compliance (Boldt and Backer, 1999; Ruff, 2001b). Given the recent advances in closed-circuit television technology, this may become a viable option, but the sensors and mountings of a camera system must be robust enough to operate in an environment prone to vibration and rock impact as well as wet, foggy, and dusty conditions (Roberts and Corke, 2000; Ruff, 2001b). As will be explored in greater depth later in this book in Chapter 9 about new technologies, the system as a whole must be intuitive to use whilst operating a large piece of machinery and must be accepted by the operators. In a review of all systems currently used, Ruff (2001b) reported that regardless of sensor type (radar, radiofrequency tags, or GPS), a video camera system would be beneficial for backup and confirmation of why the alarm was being triggered. Due to their successful implementation in above-ground applications, the mining industry has started to look at using cameras on underground machinery. The Mine Safety and Health Administration (MSHA) supports adding video cameras as a visibility aid and subsequently issued a Directorate of Technical Support Accident Reduction Program that aims to eliminate blind areas by providing guidelines for positioning and mounting cameras and monitors (MSHA, 2001). Roberts and Corke (2000) found that an LHD cab-mounted laser scanner solved the problem of not being able to mount on the front due to a loaded bucket but still allowed forward and rearward travel monitoring. Therefore, new technology has the potential to solve line-of-sight issues associated with the operation of mobile equipment; however, there are significant human-related issues that need to be addressed. These issues will be explored further in Chapter 9.

chapter eight

Controls and displays

8.1 Controls and displays: Overview

Mining equipment utilises a wide range of different controls and displays. These vary from relatively simple controls such as steering wheels and levers, which have a direct physical link to the movement or action of the equipment (and the "display" may be considered to be the direct observation of the resulting movement), to systems in which the operation of a control is associated with a response which is not directly perceptible by the operator and information about the resulting state of the equipment is provided by a display. This might range from a simple display such as a temperature or pressure gauge, to video displays used in systems such as the teleoperation of equipment located at a distance from the operator, or a display provided by a computer screen. In these latter cases, the information provided to the operator through a video display or other display is crucial to allow the operator to achieve accurate control of the equipment and work process. Advances in sensor technology allow the possibility of such displays to include "augmented reality" in which information which would not be available to an in situ operator.

An example of this is a telerobotic rock breaker developed by the Commonwealth Scientific and Industrial Research Organisation (CSIRO) in Australia.* Integrated augmented reality and virtual reality technologies are being combined with video feeds. Operators can therefore see additional information such as virtual representations of the rocks and the ore bins (Figure 8.1A and 8.1B).

Regardless of the specific nature of the controls and displays, human factors principles exist which should be considered during their design and deployment.

8.2 Control design principles

General issues for the design of controls include the choice of control type (e.g., lever, knob, pedal, joystick, or voice), resistance, control order, coding (e.g., location, shape, or length), and directional compatibility. A

* See www.csiro.au/science/Mine-control.html (CSIRO, 2009).

(A)

(B)

Figure 8.1 (A–B) Teleoperated rock breaker with augmented reality displays (Courtesy of CSIRO, Australia).

wide variety of physical devices with different features and properties are available for use as controls. The appropriate control choice logically depends on the function of the control. For example, one fundamental control characteristic is whether the control is discrete or continuous. Other characteristics include whether the control is linear or rotary, unidimensional or multidimensional (e.g., a joystick), and dynamic or static. The features and the choice of control type are well covered by standards such as AS4024.1903 (Standards Australia, 2006), MIL-STD-1472F (US Department of Defense, 1999), and NASA-STD-3000 (NASA, 1995).

8.2.1 Control resistance

Other critical features of the control design include the degree of resistance to movement (for dynamic controls) and the type of resistance (e.g., elastic, frictional, or viscous). For example, for a control with elastic resistance (such as a spring-loaded control), the resistance to movement increases as the control is displaced from a neutral position, providing feedback to the operator about the amount of displacement. In contrast, for a control with viscous resistance, the resistance to movement is proportional to the speed of the movement. This form of resistance reduces high accelerations and promotes smooth movements, something which might be important if, for example, the control is operated from a vehicle moving over rough terrain.

8.2.2 Control sensitivity

Another factor which influences operator accuracy during a tracking task (such as steering a vehicle) is control sensitivity, or gain, sometimes expressed as a control–display ratio. An insensitive (low-gain) system in which large movements of a control are required to cause small responses (high control–display ratio) is effortful, in that large control movements are required. A sensitive or high-gain system (low control–display ratio) is difficult to control accurately because small control errors are amplified. An optimal control–display ratio (gain) will exist for any system, although the optimal gain may differ across operators as a function of experience and skill, with skilled operators performing better with higher gain. This might suggest that an adjustable gain (such as pointing device sensitivity) may be desirable in some circumstances (e.g., where operator expertise varies widely, changing either across different operators or with practise).

8.2.3 Control order

The control order also influences operator performance. A "zero-order" control is one in which a direct relationship exists between the state of the control and the displacement of the response. For example, the relationship between the angle of a steering wheel and the angle of a vehicle's wheels is zero order. To return the vehicle's wheels to straight ahead, the driver need only return the steering wheel to the neutral position.

A first-order control is one in which the state of the control is linked to the velocity of the response. The relationship between steering wheel angle and vehicle heading is a first-order relationship (when the vehicle's speed is constant). For as long as the steering wheel maintains a constant angle from the neutral position, the heading of the vehicle will continue to change with constant (angular) velocity. If the steering wheel is returned to neutral, the vehicle will continue in a straight line, but with a new heading.

A second-order control is one in which the state of the control causes acceleration of the equipment response. The relationship between steering wheel and lateral position on the road is a second-order relationship. For example, to achieve a lane change, a driver first alters the angle of the steering wheel to one side, causing a constant velocity change of heading, but an accelerating rate of lateral movement. When the desired degree of lateral movement has been achieved, the driver must rotate the steering wheel away from neutral in the opposite direction, and then return to neutral when the original heading is regained. Higher order controls are generally more difficult to operate, and a general principle is to utilise lower order controls where possible.

8.3 *Reducing control errors: Guarding, feedback, mode errors, coding, and directional control–response relationships*

One aim of the application of human factors to equipment design is to reduce the probability of human error, and examination of information about errors which have been made can provide valuable information to improve equipment design. Injury narratives reveal at least three relatively common categories of control errors which have potential to cause injury to the operator or another person: (1) inadvertent control operation, (2) operation of the incorrect control (a selection error or mode error), or (3) operation of the correct control but in the wrong direction (a direction error) (Burgess-Limerick and Steiner, 2006, 2007). Other consequences of sub-optimal control design may include control operation delays (increased reaction time and/or movement time), decreased accuracy of response (for example, in a tracking task such as steering), and increased training time. Some of these consequences are linked in that a speed–accuracy trade-off is observed. That is, operators can reduce speed to increase accuracy, or vice versa. Optimal control design will allow both speed and accuracy to be maximised.

8.3.1 *Inadvertent control operation*

The prevention of inadvertent operation of controls is an important consideration in the overall design of equipment. Location of controls where they are unlikely to be inadvertently contacted is one option, as is guarding (e.g., Figure 8.2A–C). This is especially important in underground environments where falling objects or material may cause control activation. It is important to ensure than control guarding does not hinder operator access, however, and this is something which should be addressed during user testing.

Another option is to require two-handed operation of high-risk control functions. The disadvantage of this strategy for preventing inadvertent control operation is that operators will find ways of defeating the control if two-handed operation is inconvenient or slower. In this case, the source of the inconvenience associated with two-handed control must be determined and designed out.

8.3.2 Mode errors

There is a temptation to reduce the space required for controls by allocating multiple functions to a single control, with the function selection controlled by a separate mode switch. The consequence of such an arrangement is the introduction of the possibility of "mode errors," that is, an error in which an unintended function is activation (a form of selection error).

For example, Figure 8.3 illustrates an underground coal continuous-miner remote control. The remote features a Shift control (lower left) which acts as a mode control to change the effect of other controls. Whilst an effective strategy for reducing the size of the remote control (an important issue), the possibility for mode errors to occur is created by this design.

8.3.3 Operation of incorrect controls

A relatively well-studied example of control selection errors occurs during the operation of underground coal-bolting and drilling equipment. The introduction of roof and rib bolting to prevent rockfalls in underground coal mines was a major safety advance (Mark, 2002). However, additional hazards were introduced in the process. Analyses of narratives describing injuries occurring in both Australia and the United States have highlighted the potential of control errors during drilling and bolting to cause serious injuries (Burgess-Limerick and Steiner, 2006, 2007). Examples of typical bolting controls are illustrated in Figure 8.4A–B.

Standardising the controls on bolting machines has been suggested many times as a means of reducing the probability of control errors. Miller and McLellan (1973) commented on the "obvious need" to redesign roof-bolting machines, suggesting, for example, that of 759 bolting machine–related injuries, 72 involved operating the wrong control, whilst Helander et al. (1983) determined that 5 percent of bolting machine accidents were caused by control activation errors. Helander et al. (1980) suggested that

(A)

(B)

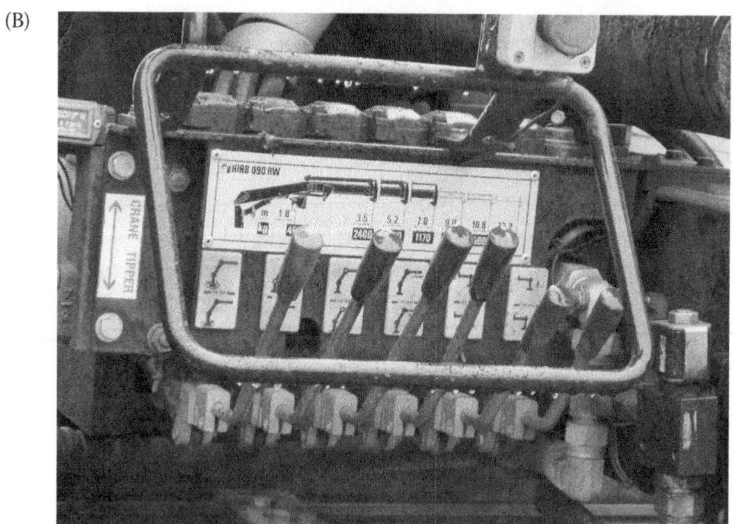

Figure 8.2 (A–C) Guarding of controls to prevent inadvertent operation.

(C)

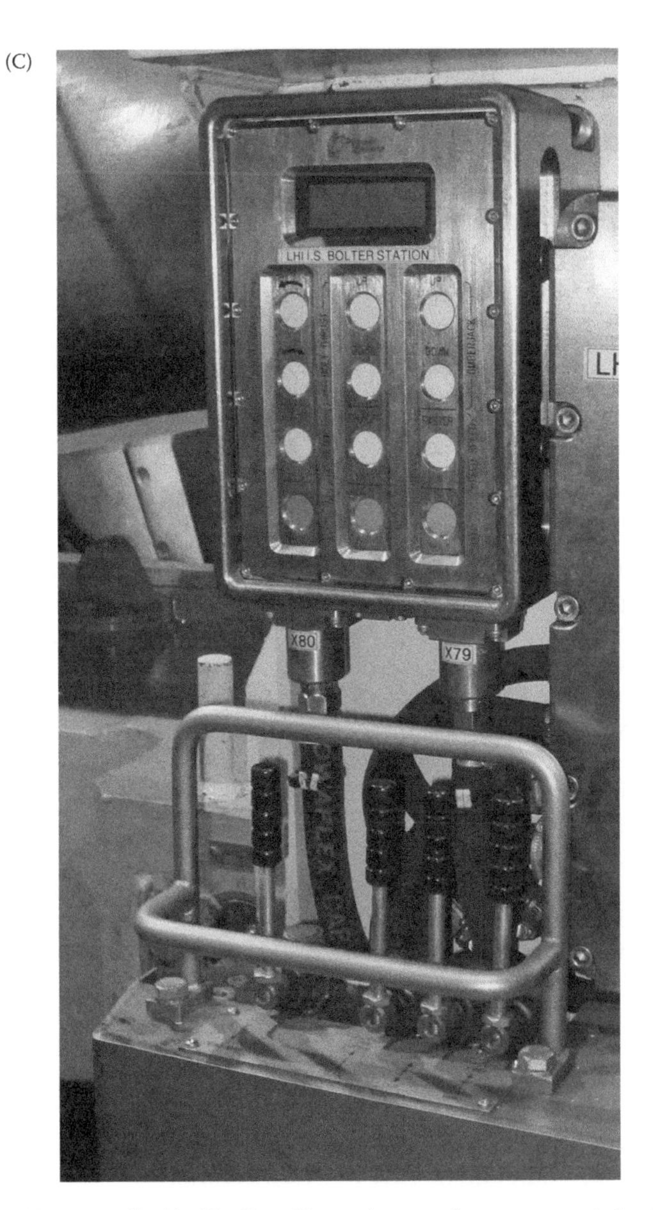

Figure 8.2 (Continued) (A–C) Guarding of controls to prevent inadvertent operation.

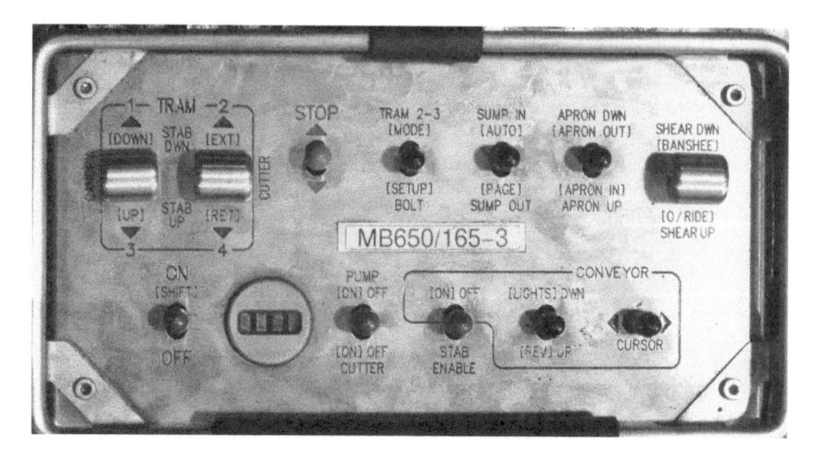

Figure 8.3 Underground coal continuous-miner remote control.

"poor human factors principles in the design and placement of controls and inappropriately designed workstations contribute to a large percentage of the reported injuries" (p. 18).

A Society of Automotive Engineers (SAE) standard that addressed these issues, titled "Human Factors Design Guidelines for Mobile Underground Mining Equipment," was defeated at a ballot in 1984 (Gilbert, 1990). A subsequent report (Klishis et al., 1993) again noted the lack of standardisation of bolting machine controls, even amongst machines from the same manufacturer, and commented on the potential for injuries due to incorrect control operation.

In a six-week period in 1994, three operators of roof-bolting machines in the United States were killed. Two were crushed between the drill head and machine frame whilst rib bolting, and the third was crushed between the drill head and canopy. A Coal Mine Safety and Health Roof-Bolting-Machine Committee was formed by the US Mine Safety and Health Administration (MSHA) to investigate, and a report was released (MSHA, 1994) which determined the causes to be the unintentional operation of controls. Amongst other suggestions, a recommendation included in this report was to "[p]rovide industry-wide accepted distinct and consistent knob shapes," although no rule was forthcoming.

Industry & Innovation NSW publishes Mining Design Guidelines to assist mining companies and equipment manufacturers meet their obligations to provide "fit-for-purpose" equipment. Mining Design Guideline 35.1 (MDG 35.1), *Guideline for Bolting and Drilling Plant in Mines: Part 1: Bolting Plant for Strata Support in Underground Coal Mines* (I & I NSW, 2010), has been published which stipulates a standard set of knob shapes for the primary bolting controls (rotation, feed, and timber jack) (see Figure 8.5A–B).

(A)

(B)

Figure 8.4 (A–B) Typical bolting controls.

Whilst shape coding is very commonly recommended within human factors texts, empirical evidence of the effectiveness of shape coding in reducing selection errors is scant. The most frequently quoted example of the effectiveness of shape coding is the instigation in World War II of the use of a wheel-shaped knob to correspond to the wheel raising

(A)

(B)

Figure 8.5 (A–B) Examples of bolting controls incorporating shape coding as required by MDG 35.1.

and lowering in an aircraft to reduce the probability of pilots retracting the wheels on landing, rather than the flaps as intended. This evidence remains anecdotal, and as the investigator concerned described the situation, shape coding was not the only control employed:

> I was asked to figure out why pilots and copilots frequently retracted the landing gear instead of the landing flaps after landing.... What I found on inspecting the cockpits ... was two identical toggle switches side by side, one for the landing gear, the other for the flaps. Given the stress of landing after a combat mission, it is understandable how they could have been easily confused.... The ad hoc remedies proposed at the time (separate the controls and / or shape code them) were substantiated in the human factors literature years later. Another remedy was a more mechanical one—installing a sensor on the landing struts that detected whether they were compressed by the weight of the aircraft. If so, a circuit deactivated the landing gear control in the cockpit. (Chapanis, 1999, pp. 15–16)

Despite the claim that shape coding was later substantiated, only two published papers exist which address the issue. Both were conceived with aviation applications in mind, and neither provided unequivocal evidence for the benefits of shape coding. This lack of evidence was commented on by Roscoe (1980), who noted,

> The discriminability of shape-coded switch knobs had been studied vigorously following World War II.... However, further application of shape coding was stalled because no investigator had demonstrated a reliable improvement in any critical switching operation attributable to the application of discriminably shaped switch knobs. (p. 274)

More recently, a series of experiments were undertaken to address the benefits of shape coding (Burgess-Limerick, 2009; Burgess-Limerick et al, 2010a). These experiments did not provide any evidence that arbitrary shape coding, such as that stipulated by MDG 35.1, was effective in reducing the probability of a selection error during the performance of a task involving selecting one of four levers in response to visual stimuli in situations in which the layout of the controls remains constant. However, for the block of trials immediately following a change of side of the controls, a significantly lower selection error rate was found for participants assigned to the shape-coded conditions when compared to participants assigned to non-shape-coded conditions.

The previous literature provides consistent evidence. Figure 8.6 is redrawn from Weitz (1947) and presents the average number of selection errors for four groups of twenty-five participants who performed 16 one-minute trials in which four of seven controls were manipulated in response to four pairs of stimulus lights. The levers were either shape coded (groups I and II) or identical (groups III and IV), and for two groups (I and III) the lever locations were changed after eight trials, whilst the lever locations remained constant for the remaining groups (II and IV). The average error data for groups II and IV demonstrate no evidence of an advantage of shape coding when the location of the levers remained constant. However, data from groups I and III demonstrate a reduction in error rates achieved by shape coding in the situation in which the location of levers is altered.

These data, in conjunction with the data provided by Burgess-Limerick (2009), support a recommendation that shape coding should be employed for bolting control levers. However, for maximum benefit, and perhaps for any benefit to be realised, the shapes must have a consistent relationship to lever function between different bolting equipment, as well as within the same piece of equipment. Consequently, the shapes proposed for bolting equipment in MDG 35.1 should be employed universally for all underground drilling and bolting equipment.

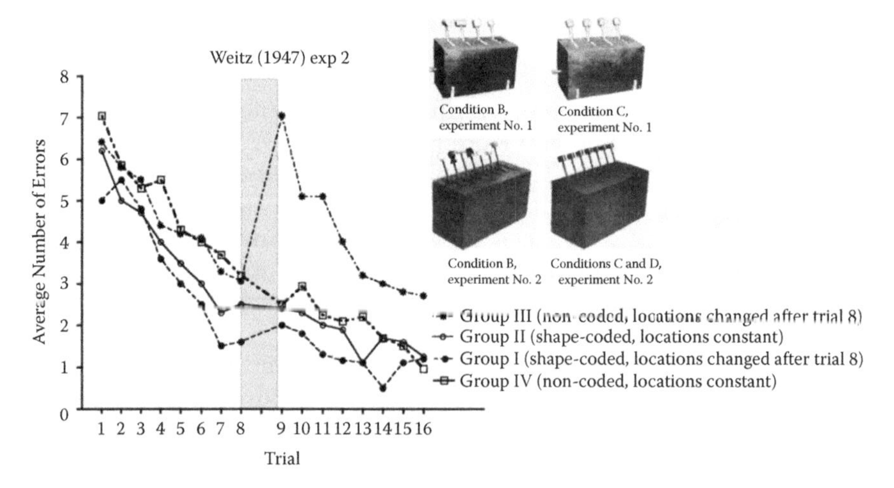

Figure 8.6 Data redrawn from Weitz (1947) from an experiment in which shape coding was shown to be advantageous only when the order of controls was altered.

Further, the provision of shape coding creates an additional risk that selection errors may be provoked if the shapes were inadvertently swapped during equipment maintenance. Consequently, it is recommended that standards which specify shape coding should also include a provision stipulating that the design of the controls must incorporate a means of ensuring that the shaped handles cannot be inadvertently fitted to the incorrect lever. This is consistent with MIL-STD-1472F (US Department of Defense, 1999, section 5.4.1.4.4.e): "Shape-coded knobs and handles shall be positively and non-reversibly attached to their shafts to preclude incorrect attachment when replacement is required" (p. 57).

8.3.4 Direction errors

The final error type considered here occurs when the correct control is operated, but the control response is the opposite direction to that intended (a directional error). A potential contributor to these errors is equipment in which the directional control–response relationships are not "compatible." As one example, consider the shuttle car illustrated below, which is typical of those used in Australian underground coal mines. These vehicles are driven using a steering wheel located to one side, and between, the two facing seats, attached to the inside wall of the cab adjacent to the shuttle car's conveyor. Two facing seats allow the driver to change seats with each change of direction and always face the direction of travel.

When the car is full of coal, the driver will move to the left-hand seat of the cab illustrated in Figure 8.7, and drive the car to the boot end, where the coal is unloaded. When driving away from the face in this shuttle car, if the driver wishes to turn the car to the right, he rotates the steering wheel clockwise, pushing the top of the steering away from the body. To turn left, the driver pulls the top of the steering wheel back towards his body, an anti-clockwise rotation. This relationship between control direction and vehicle response is "compatible" because it corresponds to the driver's natural expectations, and is typically executed without error or delay.

However, when the driver has emptied the car at the boot end, and changes seat to return to the face, the task is more difficult. The driver is now sitting in the right-hand seat seen in Figure 8.8. Now, if the driver wishes to execute a right-hand turn, he must push the top of the steering wheel away from himself (anti-clockwise rotation), and to turn left he must pull the top of the steering wheel towards his body (a clockwise rotation). This is an "incompatible" control–response relationship.

The consequences of incompatible control–response relationships generally are an increase in errors, and/or a reduction in the speed with

Figure 8.7 Shuttle car cab with driver seated facing outby.

Figure 8.8 Shuttle car with driver seated facing inby.

which tasks can be performed, and this conclusion has been confirmed in a series of experiments in a virtual reality simulation of the situation (Zupanc et al., 2007; Zupanc, 2008).

One solution to this problem is to provide a single rotating seat and a steering mechanism which remains always compatible such as that provided in the "Ergocab" shuttle car illustrated in Figure 8.9A–C.

Other examples exist where the design of controls requires careful consideration of directional compatibility relationships—consider once again the design of controls for underground coal-bolting equipment. Mining Design Guideline 35.1 (I & I NSW, 2010) requires, "The

(A)

(B)

(C)

Figure 8.9 (A–C) "Ergocab" shuttle car design featuring a single rotating seat and all-ways compatible steering.

direction of operation of manual controls should be consistent with the direction and response/movement of the actuator or the plant, where practicable."

The importance of ensuring "compatible" directional control–response relationships is unanimously agreed. However, determining the appropriate directional relationship in any specific circumstance is not always straightforward. It is relatively common on mining equipment to find situations in which downwards movement of the horizontal control lever causes upwards movement of the controlled element such as a boom, timber jack, or drill steel. Whilst some authors (e.g., Helander et al., 1980) have suggested that this is a violation of compatible directional control–response relationships, Simpson and Chan (1988) suggested that the response may be compatible if the operators assume a "see-saw" mental model of the situation, where moving the near end of the control downwards causes the far end (and the controlled element) to move upwards.

Simpson and Chan (1988) investigated this situation through an experiment in which 144 people reported the direction they would move a control lever to achieve a specified effect, using a one-tenth model of a drill-loading machine. The results indicated that whilst the majority of people reported responses consistent with a "see-saw" mental model, the stereotype was far from universal, and up to 33 percent of people reported expectations for "up = up." Extremely strong expectations were reported for the movement of vertical controls, however, with more than 90 percent of people expecting a backwards movement of a vertical lever to cause an upwards movement of a controlled element.

These expectations are not consistent with previous results, however (see Loveless, 1962, for a comprehensive review). For example, Vince (1945, as cited by Mitchell and Vince, 1951) reported that participants' expectations are for an upwards movement of a linear control to result in an upwards linear movement of an associated display. This principle might be called the "principle of consistent direction" and is generally reflected in current standards such as MDG 35.1. Vince and Mitchell (1946) were similarly reported to have examined relationships between linear movements of control and displays in different planes, finding that a forwards movement of a vertical control placed in front of participants was expected to cause an upwards movement of an associated linear display.

Of relevance to the design of bolting controls, Humphries (1958) noted that directional expectations were influenced by operator position with respect to the control and displays. Participants were reported to expect a control movement to the right of the body to produce a display movement to the right of the field of view, and for a control movement away from the body to produce an upwards movement of the display.

More systematic investigations of the effect of operator orientation with respect to the display were undertaken by Worringham and Beringer (1989, 1998). The general principle of consistent direction was extended to accommodate situations in which an operator uses a control located to one side, or behind, whilst looking straight ahead. In this case (and consistent with Humphries, 1958), the compatible directional relationships were reported to be ones in which the movement direction of the control in the virtual visual field (as if the participant was looking at the control) was consistent with the movement of the controlled element. This principle is referred to as "visual field compatibility," that is, "[T]he direction of motion of a control, as defined by its movement when viewing the control, should correspond with the movement of the controlled object in the operator's visual field, whilst viewing the object" (Worringham, 2003, p. 14).

More recently, the issue has been examined in a series of experiments involving both virtual and physical simulations (Burgess-Limerick, 2009; Burgess-Limerick et al., 2010b). The apparatus illustrated in Figures 8.10 (A–B) and 8.11 (A–B) was used in a series of experiments in which the orientation of the control levers was varied between horizontal and vertical; the location of the levers was varied between left, front, and right; and the directional relationships between the lever movement and the response of the physical or virtual model were altered.

The experimental paradigm was effective in discriminating differences in directional error rates and reaction times between directional control–response relationships, and the results from both experiments were largely consistent. With few exceptions, the results confirmed the general applicability of the principles of consistent direction, and visual field compatibility principle (Worringham and Beringer, 1998). In particular, the finding that directional error rates were minimised when upwards movements of a horizontal lever caused upwards movements of the controlled device was consistent with the data reported by Mitchell and Vince (1951) and not with the participant expectations reported by Simpson and Chan (1988). This discrepancy raises the possibility that self-reported directional expectations are not necessarily predictive of behaviour, or of the ease of learning different directional control–response relationships. Hoffmann (1997) and Chan and Chan (2003) have similarly reported discrepancies between reported directional expectations and actual behaviour.

The control of slew left and right was associated with a relatively high probability of control errors (median error rates above 2 percent) in most of the situations examined. The exceptions were situations in which a vertical lever was located to a participant's right or left with a directional control–response relationship such that moving the lever away caused the device to slew in the same direction. That is, a vertical lever located to a participant's right was paired with a directional relationship such that

(A)

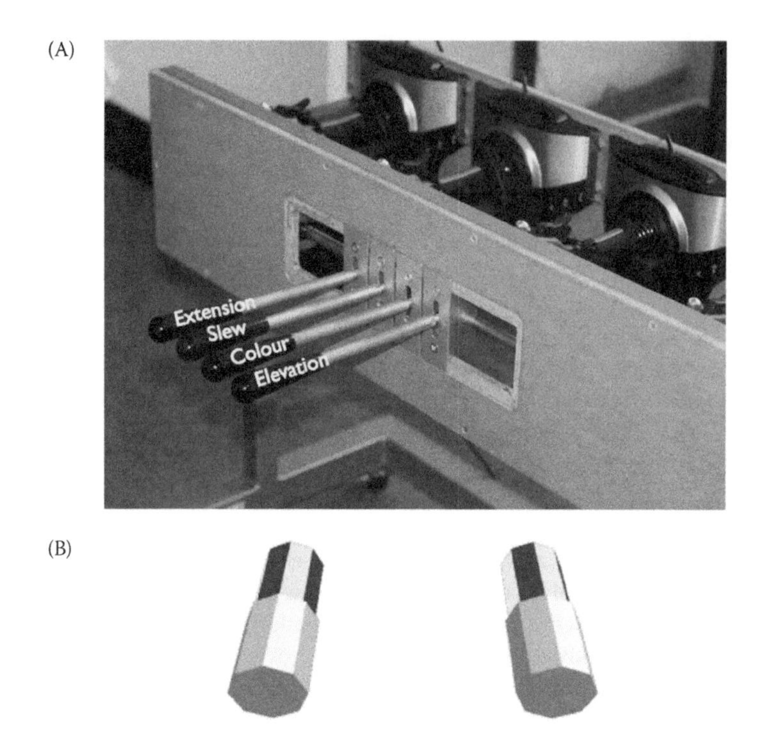

(B)

Figure 8.10 (A–B) Virtual reality simulation of a generic device used to assess directional control–response compatibility (Courtesy of The University of Queensland).

pushing the lever away caused a slew to the right and was associated with very few direction errors, and similarly very few direction errors occurred when a vertical lever located to the participant's left was paired with a directional relationship such that pushing the lever away caused a slew to the left. This outcome is consistent with the principle of consistent direction. Directional error rates were higher when the direction of movement of the slew (left or right) was perpendicular to that of the control (i.e., all front on situations examined, and all horizontal lever orientations) regardless of the directional control–response relationship. These situations should, therefore, be avoided.

In contrast, the optimal directional control–response relationship for extension or retraction was not always consistent with the principle of consistent direction. Whether the controls were located in front or to either side of the participants, and regardless of whether the extension or retraction occurred when the virtual device was vertical or horizontal, directional error rates were significantly lower in the control–response conditions in which raising a horizontal control, or pushing a vertical control away, caused extension of the virtual device. This finding suggests

(A)

(B)

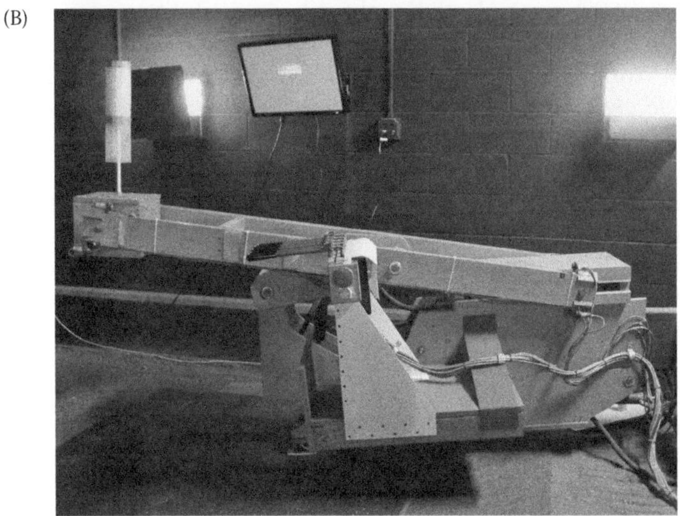

Figure 8.11 (A–B) Physical model of a single boom bolter arm used to assess directional control–response compatibility (courtesy of National Institute for Occupational Safety and Health, Office of Mine Safety and Health Research [NIOSH]).

that a compatibility relationship between "lengthening or shortening" or "extension or retraction" exists in addition to the directional movement relationship.

That said, the lowest error rates occurred when these movements were also consistent with the principle of consistent direction (e.g., when a pushing a vertical lever located on a participant's right caused horizontal extension to the right), suggesting that the effects are additive.

The compatibility of the directional control–response relationship for the situation in which the controlled device was elevating or depressing (via rotation either towards or away from the participant; or via clockwise or anti-clockwise rotation) depends on the orientation of the control. When the control was oriented vertically on the left or the right, the principle of consistent direction holds, in that very few directional errors occurred for a vertical control in front of the participant when pulling the lever back caused the controlled device to rotate in the same direction (CRR2). Similarly, fewer direction errors occurred for clockwise and anti-clockwise device movement when the corresponding movements of vertical controls located to a participant's left or right were in the same direction.

In the situation where a vertical control to the left or right was used to elevate or depress the device directly towards or away from the operator, a directional control response in which pulling the lever towards the operator caused elevation resulted in fewer errors. This relationship is consistent with the results reported by Humphries (1958) and Worringham and Beringer's (1998) principle of visual field compatibility.

When a vertical control in front of the participant, moving towards or away from the participant, was used to control clockwise or anti-clockwise elevation and depression in the frontal plane, there was no advantage of either directional control–response relationship, and the rate of direction error was always relatively high. In this situation, neither directional relationship was compatible, and this situation should be avoided. Where horizontal controls were used to cause elevation, either towards the participant or in a perpendicular plane, fewer direction errors occurred in situations in which an upwards movement of the lever caused elevation.

Consequently, the following recommendations for the design of bolting rigs are justified: (1) where horizontal levers are used to control the extension of a timber jack to either the roof or rib, an upwards movement of a horizontal control should be employed to cause extension; and (2) where horizontal levers are used to control drill feed to roof or rib, an upwards movement of a horizontal control should be employed to cause feed.

Similarly, when extension was controlled by a vertical lever, fewer directional errors occurred without exception when lengthening of the device was caused by a movement of the vertical lever away from the operator. Consequently, the following recommendations for the design of bolting rigs are justified: (3) where vertical levers are used to control extension of a timber jack to roof or rib, a movement of the vertical lever away from the operator should be employed to cause extension; and (4) where vertical levers are used to control drill feed to roof or rib, a movement of the vertical lever away from the operator should be employed to cause feed.

In general, the lowest error rates for extension occurred when raising a horizontal lever, or pushing away a vertical lever, to cause extension was also consistent with the principle of common control–response direction, for example pushing a vertical lever located to a participant's left to raise extension to the left, or raising a horizontal lever to cause extension vertically. An optimal design of bolting rigs would ensure that the maximally compatible relationships between control and response are always maintained.

These recommendations apply to the design of new bolting rigs. A variety of directional control–response relationships are found on bolting rigs currently in use. The consequences of making alterations to those existing rigs which do not conform to the above directional control–response relationships are unknown. It may be that for operators who are accustomed to operating controls which have directional control–response relationships which differ from those recommended here, a change may increase the probability of a directional error, at least in the short term, and changes to existing equipment may consequently be undesirable. Alternately, it may be that the changes would be accommodated without difficulty and that changing existing equipment is desirable.

Whilst the above results were motivated by specific design issues associated with bolting machines, the results are likely to have general applicability, given that similar results were obtained using both a physical model of the single boom bolter, and a virtual simulation of a generic device which was not representative of any particular piece of equipment.

8.4 Display principles

8.4.1 The importance of visual information

Vision is the dominant perceptual modality. This does not imply that other sensory information is not used. Auditory information (sound) is sometimes also of key importance: for example, the sound of a siren can provide vital information to be acted upon, often in situations where vision alone could not do so. Or operators may hear sounds which indicate that the equipment requires maintenance, whereas they could not see such information without stopping the work process and specifically checking for it. Likewise, a maintainer may detect smells that provide information about the engine state. In addition, touch and proprioception (sense of body position) are used, for example, to inform operators whether their feet are on the equipment's foot controls, so that they do not have to look down to their feet each time they want to change their position. Human vestibular apparatus (sense of balance) can help to indicate that the equipment is on a slope, for example. Of course, operators' vision can often tell them this, but

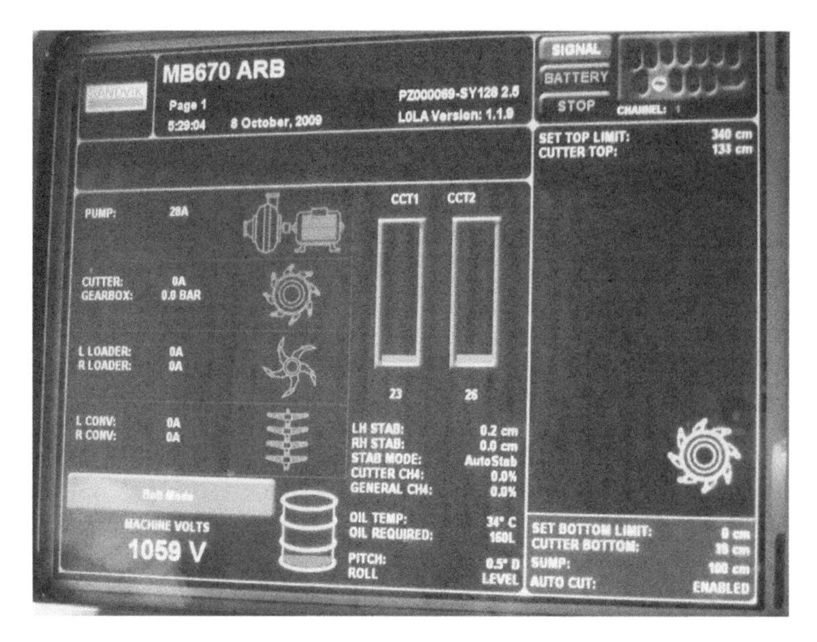

Figure 8.12 Example of a display with both static and dynamic features.

the vestibular apparatus can add to the visual information, particularly in situations where vision is diminished (such as at night) or even obscured.

8.4.2 *Types of visual information*

However, given the importance of vision, most mining equipment displays are visual. Within this, a frequently used distinction for displays is between dynamic and static information. Dynamic information can include fuel monitors, stop-go lights, and speed–engine gauges on mining mobile equipment. Warning labels are static information. Computer text and graphics have traditionally been thought of as static information; however, the increasing use of display screens to show data parameters is blurring this distinction (e.g., Figure 8.12).

8.4.3 *Warnings and alarms*

In addition to the *type* of displayed information in mining equipment, another way to categorise is to consider the *function* of the display. In addition to displays that show "routine" information, another category of displays deals with emergency or other non-routine equipment-related information. Both audible warnings (usually known as *alarms*) and visual ones will be briefly considered. Other warning modalities exist (e.g., by smell, or movement or vibration), but coverage here will be confined to the auditory and visual versions as these are likely to be the most used for many years to come in mining equipment.

There are several main objectives that audible or visual warnings must accomplish: they must attract attention (i.e., they must be noticed or heard), they must be attended to by operators and maintainers, they must be understood by the users, and they must incline the operator to follow, heed, or comply with them. All that seems straightforward, but in many complex mining systems there are a large number of unique alarm or warning systems. For example, in mineral process control rooms it is not uncommon to have 100 or more different auditory alarms and visual warnings. Even the average light vehicle can have over twenty-five different warning lights or alarms (including engine parameters or seat belt reminders). Apart from the large number of alarms, often these can be too loud and sound too frequently. In emergency or abnormal situations, it is possible that many of these alarms and warnings might be activated at the same time, which can potentially overwhelm an operator. Under such conditions, the operator may adopt an inappropriate alarm-sampling strategy, or may just ignore them.

As with many other mining equipment displays, designing effective alarms and warning essentially involves the following:

- Understanding the equipment-related tasks performed by operators, and the environment in which the visual warning would be displayed
- Using good human factors guidelines (e.g., for alarms, loudness or intensity, pitch, frequency timing, and duration are all important to consider; likewise, for visual warnings, their position, size, contrast, and intensity are vital)
- Iterative design and testing with the relevant operator population

8.4.4 Key display design principles

Considerable research attention has been given to the design of displays, leading to the development of well-established principles. Wickens et al. (1998) provide a good summary; the principles include the following:

- Avoid requiring absolute judgments (relative judgments are less error prone).
- Utilise redundancy (simultaneous coding of information in different physical dimensions).
- Utilise pictorial realism (the display resembles the information displayed).

Table 8.1 EMESRT Controls and Displays Design Philosophy

	Controls and Displays
Objective	The objective is to minimise risk of operator error related to the understanding or use of controls and displays (including labelling) to as low as reasonably practisable (ALARP), including consideration in design for foreseeable human error.
General outcome	The intended design outcome should include the following: Controls and displays that are appropriately located, intuitive to use, consistent, and fail-safe. Warnings and alarms that are designed to be detectable, unambiguous, simple, and meaningful. If multiple alarms are possible, then they should be tested and integrated so that they minimise the risk of overloading an operator. Labels that are durable, clear, and appropriately positioned. In addition, the intended design outcome should minimise injury from contact with controls and displays in the event of a collision or accident.
Risks to be mitigated	1. Risk of incorrect use of equipment controls, due to the following reasons: a. Not fully understood or misunderstood b. Not easily reachable (particularly frequently used and/or safety-critical ones) c. Not consistent with other controls d. Not matching the associated displays (and compatible in both their location and movement direction) e. Not appropriately considering simultaneous control operation f. Unintentionally operated g. Incorrectly selected h. In an unexpected operating mode (mode errors) 2. Risk of visual displays being illegible (e.g., unsuitable font size or colour contrast), not visible from the operator's position, in all conditions, or incomprehensible (e.g., auxiliary equipment interfering with visibility of primary displays) 3. Risk of warnings and alarms: a. Not being seen, heard, or understood. b. Not being reliable or sufficiently sensitive. c. Being overused, ignored, and compromised. Where multiple warnings and alarms might be activated, a method of integrating or prioritising them should be used, such as matching their loudness or brightness to the level of warning criticality.

	4. Risk of incorrect or inaccurate calibration or maintenance of displays, warnings, or alarms 5. Risk of ineffective labels due to the following reasons: a. Not being durable, clear, readable, and understandable b. Not using standardised terminology c. Not being positioned appropriately with the control or hazard 6. Risk of symbols used in labels, displays, and warnings not being fully understood or being misunderstood 7. Risk of injury from controls:
Examples of industry attempts to mitigate risks	a. Indicators, gauges, metres, and lights monitoring the condition of the equipment are clearly visible from the operator's position, day or night (e.g., backlit and adjustable). b. Instruments, controls ,and other gauges that are marked in colour-coded metric units; gauges with a single scale with no multipliers. c. Self-test facility to test warning lights. d. Instrument panels made up of a number of easily removable modules to enable various instruments and the wiring loom to be easily serviced. e. Labels designed and fitted to be permanent, durable, and readable, fitted to non-removable parts where possible (e.g., frame). f. Labelling of location of emergency egress points and emergency stop. g. Designated isolation points clearly labelled to identify the system they control. h. Where isolation points are located behind engine, or other covers, clear isolation labels located both beside the isolation point and on the outside of the cover. i. An indicator to display when an engine shutdown timer is activated. The "off" switch and indicating light mounted (and labelled) on the front dash of the cabin. j. Label showing all limits of application (such as maximum vehicle height). k. Warning and danger signs—labelling for all specific hazards and tasks (e.g., noise). l. Designated towing, jacking, and supporting locations labelled. m. Making control and displays low profile, secure, and smooth to minimise injury risk in the case of an accident.

- Minimise the effort required to access information (most frequently required information is the most easily accessed).
- Proximity compatibility (information which requires integration should be located in close proximity).
- Predictive aiding (providing information about what will happen in the future rather than the current state is beneficial, especially for higher order control systems).
- Consistency (errors are likely to be reduced if the meaning of displays is consistent across different equipment).

8.5 Case Study: The EMESRT controls and displays design philosophy

The recent work of the Earth Moving Equipment Safety Round Table (EMESRT) was introduced in Chapter 2. EMESRT produces "design philosophies" for key aspects of equipment design. The latest version of the controls and displays design philosophy for surface-mining equipment is shown in Table 8.1 (two of the authors of this book had a major role in developing it, and a couple of annotations are added here to put a couple of issues into context). It shows key aspects of equipment controls and displays, and was developed for designers and equipment manufacturers as well as mine site personnel. As with other design philosophies, it shows an objective, a general outcome, a list of the key risks to be mitigated, and some examples of industry attempts to mitigate the risks. It is presented here to summarise many of the concepts introduced in this chapter.

chapter nine

Automation and new technologies

9.1 Why are new mining technologies and automation being developed and deployed?

Automated and new mining technologies are being increasingly developed and deployed for a number of efficiency and/or safety reasons. These vary by their precise application to different aspects of mining, but they generally fall into one or more of the following broad categories:

- *Removal of operators from hazardous situations.* This includes from near large mobile equipment, or even total removal from a mining method (e.g., underground room and pillar mining, in theory). This category also includes the reduction in the need for hazardous maintenance or exploration tasks.
- *Lower cost of production.* In one sense this is a catch-all category, and virtually all new or automated technologies need to achieve a positive return on investment (across the full life cycle of the equipment). Examples include more ore dug and transported, or more efficient process control operations.
- *Requirements for enhanced precision.* As will also be mentioned later, an example of this is automated blast hole drilling—where not only is there a potential safety benefit, but also the correct location of the blast holes can be more accurately achieved through automated systems. Similarly, for in-vehicle assistance systems in mobile mining equipment, this might include roadway departure warnings or better braking and speed-limiting systems (to allow the mobile equipment to be operated more precisely when in busy operations).
- *Less environmental impact.* Automation and new mining technologies can, in theory, be more sustainable, minimise the need for land reclamation (e.g., by using keyhole mining methods, rather than open-cut operations), and require less energy to extract and process the commodity.
- *Being able to mine areas previously inaccessible.* For example, being able to mine in hard-to-reach locations that previously could not be mined economically (e.g., at greater depths or lower seam heights).
- *More data and information.* Certainly, such aspects need to be handled correctly, so they are not perceived as intrusive and as unnecessarily

spying on an operator. However, the capacity to be able to collect more data, often in real time, on the performance and state of mining equipment can be of considerable advantage for several reasons: it can help with equipment maintenance scheduling, may be useful in identifying the equipment-related issues which limit performance, could help identify operators who are not performing optimally (and so, for example, might benefit from additional training on selected aspects), and can help with emergency response or in abnormal situations.

- *Reduced manning.* As will be seen in this chapter, it is a myth to think that automation fully removes the need for all human involvement. However, it would change, and in some cases reduce, the need the humans, at least those on the front line. For example, automated haul truck or train movement of ore is becoming more common, and so requires less direct involvement of haul truck operators or train drivers (at least at the location of the equipment itself).

At the time of writing, intensive research and development work in automation and new mining technologies is being undertaken by many major mining companies, equipment manufacturers, and universities. The focus of this is still slightly more on open-cut operations (in part, because technologies used in domains such as road and rail transport can be more easily adapted to open-cut than to underground mining), and some of the main mining equipment types that are being particularly investigated include haul trucks, blast hole drills, rock crushers, and ore trains. However, in future years, it seems likely that automation or new technologies in some form will be further applied across virtually all mining equipment (e.g., shovels and excavators) and mining methods.

All this rightfully sounds impressive and exciting. However, significant human factors issues remain, albeit sometimes different from the traditional concerns of the subject. There is less of a focus on manual tasks and environmental ergonomics, but more of a focus on interface design, acceptance of new technologies, and the changing skill requirements for people to operate and maintain the new equipment. Furthermore, as will be seen in this chapter, there is the potential for automated systems to overload, confuse, and distract, rather than support or assist, an operator. Therefore, general approaches mentioned elsewhere in this book—like standardisation, appropriate training and risk assessments, alarm integration, operator and manager consultation, and input and feedback by authorities and inspectors—are all vital. Equally, using human factors principles to help design the equipment, procedures, training, evaluation, and implementation of such systems is also of key importance.

9.2 Levels of automation

Of course, there is no single thing called *automation* or *new technology*. These terms cover a wide range of devices, approaches, systems, and components (Sheridan, 2002). Therefore, before looking into their possible behavioural effects, it is necessary to distinguish the major different types. There are different levels of automation operating in most industries, and although the uptake of them in mining has been often slower than in some domains (e.g., aviation), they still can be categorised in roughly similar ways. Each level poses different potential for serious human-related issues which may significantly limit the benefits gained from the technological advances.

To simplify greatly, automated systems and new technologies in mining can generally be separated into three broad categories related to *system control*:

- *Lower level automation*. This category includes systems that simply warn or inform an operator, supervisor, or maintainer. Examples here are proximity warning alarms for mobile equipment, new communication technologies (including data transfer of equipment status to another location), or warnings that maintenance of the equipment is due. As previously found in the aviation and road transport domains, well-researched human factors automation topics here include operator acceptance of these technologies, and possible distraction and information overload from systems, especially during operational tasks. In this automation category, the operator is still in full control of the equipment at all times, and the technology merely warns or assists.
- *Midlevel automation*. This may involve removing operator control at certain times but not others, or having the operator controlling the equipment from a nearby location. Examples of this may include during routine operations (e.g., a coal train or even a haul truck set at a predefined speed—and the operator is merely a passive monitor at such moments unless intervention is necessary), and line-of-sight control of underground equipment such as continuous miners. Collision detection technologies that actually automatically stop the equipment when collision is detected as imminent (rather than merely providing a warning) would also fall into this camp. In this automation category, the operator is in control of the equipment at most times, but certain functions (e.g., speed) are automatically controlled by the system and overseen by the operator.
- *Full automation*. In this category, no operator is physically present at the work location. This may include equipment control from a remote location (on the mine site, or even at a completely separate

geographical location) by means of a computer screen, joysticks, and other controls and displays. It may also include more full-system control being given over to the equipment (e.g., mobile equipment may be programmed to drive between two defined locations, and the "operator" is merely overseeing this operation in a supervisory capacity).

On the face of it, this category may have fewer human factors issues, but lessons learnt from some industries (e.g., aviation) show that this is not true. In addition to equipment testing and calibration, setup, routine, and emergency maintenance, there is the control of equipment during emergencies or abnormal states. Similarly, there are also the issues of acceptability of automation to operators, loss of situation awareness, boredom associated with what has become a vigilance task, deskilling, and operator behavioural changes (short or long term) with regard to different degrees of automated systems, and how this impacts upon risk. Whilst this latter point is relevant to some degree to all levels of automation, it is particularly acute for full automation where the degree of system control by the operator is less.

9.3 The importance of considering human–machine interaction in automated mining equipment

Most mining tasks involve operators or maintainers interacting with some kind of technology. However, as noted above (and in Chapter 8), the technology being interacted with is becoming increasingly complex. The trend seems likely to continue, and the use of automation or new technologies in mining will undoubtedly become even more widespread.

9.3.1 Why consider the human?

As ergonomists and human factors specialists, we argue that optimising the technologies (or, more specifically, the human–machine interfaces of these technologies) so that they match the individual operator's skills, capabilities, and limitations, as well as better integrating those technologies within the systems of work, are key future issues to consider.

In many other domains, the rapid growth in new technologies has often seen them being deployed as the technology becomes available, without them being systematically designed and evaluated from a human-centred perspective. However, to be successful, these systems must take into account the human element: for example, they must be acceptable to the user. As found by Horberry et al. (2004) for industrial forklift trucks, operators may adapt positively or negatively to new technologies. Positive adaptation occurs when a new technology brings about a positive change in operator behaviour, such as when a new speed-limiting system saves fuel and increases safety whilst being acceptable and well liked by the operators. Negative adaptation may make the operators engage in riskier behaviours. Sadly, still little research has been undertaken into the acceptability of many new technologies in the mining domain. If operator requirements and preferences are not well understood before new systems are introduced, the systems may be unacceptable when deployed. Technologies that are not accepted by operators are less likely to be used properly and are more likely to be sabotaged or misused; thus, any inherent potential for increasing safety or efficiency may not be fully achieved.

Another fundamental reason is that unless the interface of the new technology or automated system is ergonomically designed, the information it presents may overload, distract, or even confuse the operator. For example, many control systems in the mineral process control industry are not being used to their full potential, and the shortcomings of the human–machine interface are one of the principal reasons for this (Thwaites, 2008). This does not necessarily imply that the human operator is flawed, but rather that the interface itself can be improved. Similarly, as will be seen in the next section, introducing automation to replace operator weaknesses can often compromise system safety rather than enhance it.

Therefore, although automation can have positive effects on both safety and productivity in mining, systematic consideration of the human element in such systems is still vital for the technologies to be optimised, or at the very least improved.

9.3.2 Approaches and lessons from other domains

To put the whole issue into perspective, it is important to relate it to work that has been undertaken in other industries, in particular road transport, process control, and aviation. Clearly, the mining environment is vastly different in many respects, but the lessons learnt from these industries, in terms of operator behavioural responses to increased automation, can be used as a broad reference point.

In the middle of the twentieth century, systematic work regarding "allocation of function" between humans and equipment or machines commenced (Fitts, 1951). The motivation for this work was the growing complexity of equipment, and the increasing use of early automation, especially in aviation and the military. This automation often had negative outcomes; for example, some aircraft in World War II were unable to be flown safely due to inappropriate ergonomics, especially by overloading the pilot at key moments during the flight.

The allocation of function approach, in essence, makes lists of tasks that machines and equipment (including computers) are good at, and tasks that people are good at. For example, machines and equipment usually have better abilities to perform repetitive and routine tasks, whereas humans are better at complex judgments and decision making in situations where the problem is ill defined. Then, the functions and tasks of a system are divided up based on who (i.e., person or equipment) is most appropriate to undertake them.

The method does have critics (e.g., how can effective person–machine integration be achieved; Vicente, 1999), but it has been used in many domains, notably process control, manufacturing, and aviation. This approach is still partly in evidence today in design principles to help maintain operator situation awareness in automated systems, in which recommendations are made concerning automation to carry out routine actions and using humans for more complex cognitive tasks (Endsley et al., 2003). Indeed, on one level this approach is difficult to argue against: modern mining equipment can carry more ore than a person can in a bag on his or her back, whereas first aid following an accident is best administered by a person. However, as will be seen below, there are many "grey areas," especially where automation or new technologies are inappropriately designed and/or introduced.

9.3.3 Some of the "ironies" of automation

Back in the 1980s, Bainbridge (1987/1983) noted a series of "ironies of automation" about some of the problems encountered at that time with increased automation in the process control and aviation industries. These show how the automation of processes and tasks can often expand, rather than eliminate, problems with a human operator. Although Bainbridge's original scientific paper is now over twenty-five years old, many of the issues she outlined are still relevant in the mining industry today. The types of "ironies" that Bainbridge listed include the following:

- Manual control skills (longer term loss of skill by experienced operators, so when they need to intervene or take control they are not as efficient)
- Being "out of the control loop" (by the operator only passively monitoring an automated system; when intervention is necessary, he or she does not know the exact system state)
- Negative operator attitudes towards newly developed automation (often mistrust and resistance to change; this may be especially true at remote mine sites where the potential for other employment is limited)

The key aspect, according to Bainbridge, is that designers of systems often try to eliminate the human as much as possible, but still leave the operator or maintainer to do the tasks that designers cannot work out how to automate. So, in terms of allocation of function, they try to allocate as many functions as possible to the technology, and leave the human operator to mop up the remainder. Although this sounds semi-sensible, this usually leaves the operator or maintainer as a supervisory controller or problem solver—not only are these skills sometimes ill suited to human capabilities (e.g., likely vigilance decrements from monitoring a process for long periods of time) but also they are exacerbated by being "outside the control loop" (i.e., as a passive observer rather than an active operator), by negative attitudes to the technology and by longer term skills fade. Also, as will be noted in Section 9.4, introducing new technologies based simply on a "smart" piece of technology being available, rather than systematically considering the need for it, is not an ideal approach.

To summarise, the results of automation research from aviation, land transport, and process control have shown that increasing automation can have unanticipated effects upon system safety and efficiency unless the human element is systematically considered.

9.4 Automation and human factors issues

With the above in mind, this section outlines more specific human factors problems and challenges associated with new and/or automated mining equipment. Building on a classification by Grech et al. (2008) and the earlier work in other domains (e.g., Bainbridge, 1987/1983), the key issues to consider include the following:

- *Poor operator acceptance of new technologies and automation after they are introduced*. This might be evidenced by negative opinions of the new devices (e.g., not trusting the outputs of the new technology), lack of use, and even sabotage or damage to them in extreme cases. User-centred design of the technology, operator and worker consultation, understanding the exact requirements of the tasks, and an ongoing feedback process to and from management can help reduce problems with new technology acceptance.
- *Poor human factors design of equipment*. Focusing on the human–machine interface, this includes a lack of equipment usability as well as information overload and distraction issues. Workload can be upon a single operator or a full work team (e.g., all maintenance staff in a shift), and it can produce problems when too high, too low, or too uneven. Another aspect is inadequate feedback to the operator, and the effects of delays between making a control action and seeing the results of the action (i.e., time lag). More generally, many of the

controls and displays issues mentioned in Chapter 8 and throughout this book are very relevant to this point.

- *Problems with integration of multiple warnings and alarms.* The last bullet point mentioned the possibilities of overload due to poor design of equipment. However, even if a new device is, in itself, well designed, problems can occur when multiple warnings and alarms are installed. A haul truck is a good example here—in addition to the alarms and warnings in the base unit, new technologies such as collision detection systems, speed limiters, communication devices, and so on are being increasingly installed. Possible overload and distraction issues from all of these alarms and warnings, especially in abnormal operations, are distinct possibilities.
- *Lack of equipment standardisation.* In a fast-moving domain such as new technologies, equipment standardisation is often difficult, and in some instances could even stifle creativity when new products are emerging. Nonetheless, eventual standardisation is important—especially where new technologies need to be integrated into existing workstations. Lack of standardisation can result in increased errors and reduced work performance (such as reverting to stereotypical responses).
- *A new device being essentially irrelevant to the task.* Often these new technologies are designed from a "technology-centred" perspective, where new technologies and "gizmo" devices are introduced when they are available, rather than by systematically analysing the problem, defining the user needs, and carefully assessing the safety benefits they might bring (Horberry et al., 2006).
- *Inadequate operator and maintainer training and support.* Although the human factors perspective is to design the equipment "correctly" in the first place, initial training and ongoing support are still very often needed. Assessment of training needs and eventual evaluation of training effectiveness are key issues (see Chapter 11).
- *Over-reliance on the technology by operators.* Sometimes a new technology seems to be well designed and fits in well with the work system, and operators wonder how they ever previously managed without it. However, problems can arise when it is over-relied on; for example, when outputs are wrong (such as sometimes occurs in GPS navigation devices used for mobile equipment) or when the new equipment malfunctions. Deterioration of skills of the individual operators as a result of the introduction of new equipment can be a particular issue here (e.g., loss of ability to perform emergency tasks if the automation fails).
- *Lack of technology physical integration.* In some parts of mining (such as drilling equipment or process control), the base equipment is often a few decades old. New technologies and devices are often added

wherever they may fit. Indeed, retrofitting (or at least adjustment) occurs with most mining equipment to some degree. In addition to the more "cognitive" problems like overload, devices irrelevant to the task, or over-reliance, there are also physical integration issues to consider. These include reach, clearance, and visibility concerns, which might be a particular issue during equipment access and egress (e.g., during routine maintenance).

- *Organisational issues.* The changing tasks due to automation require consideration of issues such as work breaks, supervision, and how operator performance is monitored. Introducing new technology often changes the nature of the tasks to be performed, so a careful analysis of the new operational and maintenance tasks is a vital early step in ensuring that organisational issues are addressed.
- *Behavioural adaptation and risk homeostasis.* As found in other domains, the introduction of automation and new technologies can sometimes result in operators engaging in riskier behaviours in automated systems. This can be seen in all levels of automation: for example, operators "testing" mobile collision warning devices or anti-skid braking system by driving nearer to other vehicles than they might do normally, or engaging in secondary tasks (e.g., paperwork) whilst supervising the operation of full automation.
- *Being outside of the system control loop.* As Bainbridge mentioned, being "out of the control loop" can result in operators only passively monitoring an automated system, so when input or intervention is necessary they do not know the exact system state. Nowadays this is often referred to as "loss of situation awareness." Whatever the name, the issues can be real. Better human factors design (e.g., control and display arrangement), training for emergency situations, and organisational planning (e.g., to minimise fatigue) can help. Equally important are regular assessments of the system (e.g., through risk assessments, auditing, and open near-miss reporting).

9.5 Case study: Collision detection and proximity-warning systems

As hinted at in Chapter 7, collision detection and proximity-warning systems are becoming increasingly important. Partly this is because of the high percentage of incidents that somehow involve collisions—especially between mobile equipment, or between mobile equipment and people (pedestrians). This in turn is partly because there are more mobile mining vehicles, especially bigger equipment with more blind spots (Bell, 2009).

At the time of writing, collision and proximity systems of various types are subject to intensive research and development work by major equipment manufacturers, smaller companies, research institutes, and mining companies. They are also becoming increasingly important to regulators (particularly in North America and Australia); indeed, in some locations their use is being strongly encouraged (even compelled) by the appropriate safety authorities. They argue that collision and proximity detection technologies are now available and could form a valuable control, especially when used with other measures such as traffic management, barriers, and vehicle separation (Bell, 2009).

It should be noted, however, that such technologies sit fairly low on the hierarchy of control—often they are low-level warnings, or at best midlevel engineering solutions (where an automated response is made by the system to brake, shut down the machine, or perform a similar response). Higher level controls are usually more effective; even so, when such systems are robust, resilient, and reliable, they can be an additional control to add to the higher level measures. Similarly, before rushing into implementing such technologies, other approaches such as examining the original design of the equipment (e.g., the issues addressed in Chapter 2) are important.

Like the term *automation* more generally, collision detection and proximity-warning systems cover a wide variety of technologies; they differ in where, when, and how they can be used. As will be seen below, all this reveals a great many human factors challenges. Many of these challenges are good examples of the general list of issues that was given in Section 9.4.

9.5.1 Uses of collision detection and proximity-warning systems

The group of technologies, of course, covers a wide scope of uses in mining. These include the following:

- Proximity detection around remote-controlled continuous miners, or other mobile plant, underground
- Dragline and shovel detection: This can be on the dragline itself (e.g., via multiple cameras), via infrastructure-mounted sensors, or by detection from the other vehicles involved (through their own collision detection devices)
- Surface operations, especially vehicle to vehicle, vehicle to light vehicle, vehicle to people, and vehicle to infrastructure
- Locomotive speed and distance—both underground and surface
- On public roads, including lane departure warning systems for large mine-owned vehicles

As can therefore be imagined, specific systems are needed—especially to correctly gauge the required accuracy, sensor technologies, and specific human–machine interface. In this domain, no single type fits all areas.

But wherever, whenever, and however they will be used, it is important to develop key performance indicators of system effectiveness (Rasche, 2009). These might include the following:

- Failures of components (e.g., the tags used in some systems) and general device reliability
- Whether people are actually using the system (and not misusing or sabotaging it)
- Recorded warnings and evasive actions
- Before-and-after studies of incidents, near misses, or other critical events

In this vein, draft criteria that might be of assistance here, in part from those employed by Horberry et al. (2004), are that the technologies should do the following:

1. Effectively address a safety problem that was a factor in previous incidents.
2. Be technologically feasible and not require overly extensive retrofitting.
3. Not be likely to face strong opposition from operators, maintainers, or managers.
4. Be capable of being integrated with other equipment, safety systems, training regimes, and operational and maintenance procedures.
5. Have no competing "low-technology" countermeasure capable of more cost-effectively doing the same function.
6. Be reliable and produce few false alarms.

9.5.2 *Types of detection technologies*

Given the number of areas where this overall group of devices can be used, it is perhaps no surprise to learn that a wide range of sensor technologies and associated tools are being used. These include radar, wifi, cameras, radiofrequency identification (RFID), databases of static obstacles, Global Positioning Systems (GPS; and 3D mapping), and ultrasonic. Some of these work better in specific environments (especially underground); however, it is not within the scope of this book to review the technical pros and cons of these aspects; besides, there are more than enough human factors challenges of concern.

But before looking in more depth at these human factors issues, two different scenarios for the use of the technology are shown below. These are loosely based on a 2009 two-day workshop on this topic that brought together equipment developers and suppliers from around the world with mine sites that had either trialed some of the technologies or were interested in doing so (Rasche, 2009).

9.5.2.1 Example 1: Underground mining

Many sensor types will not work underground, and the intrinsic safety and resulting certification requirements associated with underground coal mines in particular create additional challenges to the introduction of technology in general, and proximity detection in particular. However, low-frequency magnetic field markers can be employed. Typically they are very accurate, and often less than 1 cm. They can be used, for example, with underground continuous miners, load haul dump trucks, roof bolters, and shuttle cars. Such systems might have different zones that have a different system response or warning; this might include an outer zone, an identification zone, a warning zone, and an automatic stop zone (where the machine is automatically turned off if a pedestrian worker enters this zone).

9.5.2.2 Example 2: Surface mining

Surface mining has an advantage over underground mining in that it can more easily build on previous work in other domains—most notably, collision detection technologies developed by the land transport and automotive domains. An example of a surface-mining system might be an RFID-based one, combined with cameras for front and rear blind spots. Detection is typically 0–100 m, but this is often configurable. Data logging might also be incorporated (especially if GPS is also used), and this can also be used to keep track of all vehicle positions on a surface mine site as well as other maintenance and production variables. Other options might include driver behaviour monitoring (e.g., fatigue), or terrain mapping. Such a system can be used for haul trucks, shovels, graders, water carriers, dozers, drills, draglines, and light vehicles.

9.5.3 Human factors issues

The human factors issues noted here are based on a recent workshop on this topic (Rasche, 2009), especially from presentations and subsequent interviews (by one of the authors) with attendees from mine sites who had trialed different technologies (in either underground or surface operations). Table 9.1 summarises the issues found (or at least anticipated) by the mines, loosely organised into different themes (with minimal rewording by the book authors).

Overall, many or most of the issues that arose are human factors considerations that have been noted in general terms earlier in this chapter. These include shorter term issues such as interface and warning design, plus longer terms challenges such as technology acceptance, reliance, trust, skill fading, unanticipated side effects, and risk

Table 9.1 A Summary of Human Factors Issues (Found or Anticipated) in Collision Detection and Proximity-Warning Systems

Change management issues
The importance of operator feedback and active involvement by site management was frequently reported as vital to achieve successful trial results.
Extra procedures and the "cost" of compliance; for example, do tags need to be carried by pedestrian workers? Changes in organisational procedures would be needed.
Overall and organisational issues
Although no direct evidence was presented, there was a fear that pedestrian workers might feel "bulletproof" when the technologies were implemented; that is, they would have a false sense of security.
The general issue of taking responsibility away from operators and instead putting it on electronics (i.e., collision technology) was voiced.
A frequently made point was to only use these technologies as an aid—to support the operator's decisions rather than removing control.
Support and technical requirements
A skilled workforce is needed to fit and maintain these systems.
Reversing cameras: should cameras be activated at all speeds? No final consensus was reached here; most likely this depends on the actual equipment and operating environment, so testing and modification might often be needed.
Suitable detection distances. For example, is 10 m enough, or can some detection be too long, so leading to false alarms and/or unnecessary information? Again, most likely this depends on the actual equipment and operating environment.
Long-term human factors issues
Long-term human factors issues of operator acceptance, sabotaging or destroying the system, and turning off alarms were raised. Ways to gain user acceptance were noted as just as important as the actual selection of technology.
Although not directly trialed, it was questioned whether operators can still drive effectively when a system malfunctions. That is, operator skills may fade over time (e.g., in hazard identification).
Side effects: Shutdowns and automated actions require careful consideration (e.g., to avoid side effects such as a vehicle stopping on a decline, thus increasing the risks).
Short-term human factors issues: How to best display information
The design of the human–machine interface was frequently reported: how to display the information to operators. Is the equipment usable? Should the technologies automatically stop the equipment or simply alarm? If they alarm or warn, should this be visual or auditory? No consensus was reached.

(Continued)

Table 9.1 A Summary of Human Factors Issues (Found or Anticipated) in Collision Detection and Proximity-Warning Systems (*Continued*)

Short-term human factors issues: How to best display information
Use a "level-of-risk" approach? Only alert when exceeded (e.g., a pedestrian enters a zone closer to the vehicle), or use levels of alerts (e.g., first a light, then a buzzer, then automated stopping or at least tone changes). Generally this latter approach was preferred.
Negative issues reported were that the technologies can overload and distract the operators (more clutter, another alarm), desensitise them (normalisation), or make them over-rely on the systems.
Type and quantity of detected objects: for example, should the system separately notify the operator of ten different pedestrian workers? Is that more important than notifying them of one other truck?
Obstructions by screens, and glare, sunlight, and dust, were raised as frequent issues.
Other uses and future developments
Data logging can be used for retrospective coaching and training of operators.
Possible simulator training of collision detection devices with operators was considered to be beneficial, and was being investigated.
Integrating multiple systems (e.g., speed limiting and collision detection) in the same piece of mobile equipment is a future challenge that mines will face.

compensation. Although not all these precise issues can be answered in this book, it is recommended that the general human factors approach of systematically analysing the tasks needing to be performed; involving operators in the device designs, evaluations, and modifications; and using human factors information to develop appropriate interfaces is of key importance to the development and deployment of successful technologies in this area.

9.6 *Mining automation and people: What can we conclude?*

So what lessons can be learnt in this area, especially for the design of future mining equipment? From the above, it can be seen that there is a long list of human factors issues that include designing the human–machine interface, minimising distraction, reducing overload, aiding operability and maintainability, improving device acceptability, managing reliance, reducing loss of skills, and identifying differing training needs.

But we certainly should not reject new technology just for the sake of it. Undoubtedly new technology will have significant efficiency benefits in mining. Our key point is that new technology needs to be developed and introduced from both a human-centred and an operational need perspective, including consideration of how it will be integrated with other existing technologies and training systems, and not purely introduced because the prototype technology is available. Involving the users at all stages of the design life cycle from concept through to the deployment and evaluation of a working system is the best way to avoid many of the pitfalls mentioned in this chapter.

chapter ten

Organisational and task factors

Although this book is about the design, maintenance, and operation of mining equipment, it is misleading to divorce the equipment from the environment in which it will be used. Earlier in this book we examined the physical environment (such as noise and temperature). The focus of this chapter is on the organisational environment—how work and tasks are organised, and how they impact mining equipment. As will be seen, such organisational and task factors can have effects on simple equipment as well as on advanced technologies.

10.1 Fatigue, shiftwork, and mining equipment

We start this process by focusing on issues related to fatigue and shiftwork. Shiftwork has long been a source of concern, and usually an economic necessity, in many work domains—with the mining domain certainly having its share of shiftwork.

Mining still is often characterised by long and irregular working hours, hence the amount of work that the operators have to perform, and the amount of rest that they are able to take, are often not optimal. A key performance-shaping factor is therefore the fatigue level of an individual operator.

It is important to separate fatigue related to people (or *operator fatigue*) from equipment fatigue that is related to, for example, stress and wear on components in a dragline or a continuous miner. As this book is about people and mining, it is operator fatigue that is of interest here.

10.1.1 What is fatigue?

To some degree, everybody knows what fatigue is, and what it means to be fatigued. However, another distinction that is often not thought about (but is often used in the scientific literature) is between muscular and general fatigue. *Muscular fatigue* comes from heavy physical work and is localised in overstressed muscles; it has been considered earlier in this book in Chapter 4 about manual tasks. Of most concern here is general fatigue. *General fatigue* can be viewed as a cumulation of all of the stresses of the day, including the duration and intensity of physical and mental work, the time of day the work is performed, and the amount of prior sleep that an operator has received; these need to be balanced by rest and recovery.

Although the issue of fatigue has been widely researched in many occupational settings, no commonly accepted definition has been established. Following Horberry et al. (2007), the term *operator fatigue* is used here as a blanket phrase that covers internal states and performance decrements associated with two broad, yet distinct, issues:

1. A need for sleep (sleepiness/drowsiness)
2. Tasks and environments that are either mentally or physically demanding (excessive task demands) or insufficiently stimulating (under-stimulation)

The importance of the distinction of sleepiness from task-induced performance effects is that the causes and countermeasures for these two phenomena are generally quite different. For example, sleepiness can be due to a lack of prior sleep or "body clock" (circadian cycle) issues, whereas task demands associated with mining or other work activities can negatively affect performance without necessarily causing sleepiness. Whereas the best countermeasure for sleepiness is obtaining good-quality sleep (or sometimes a nap), the countermeasures for other task-induced effects may include changing the physical or mental demands of the job, equipment, or task.

10.1.2 Fatigue measurement and impacts

Operator fatigue (either sleepiness or task induced) is not directly measurable. Instead, "indicators" of fatigue are employed (e.g., subjective ratings, or operator performance measures); however, these are often very difficult to identify following an accident or incident at work. Quantifying the exact number of mining accidents and incidents that involve some kind of operator fatigue as a causal factor is extremely problematic, as Cliff and Horberry (2008) found when investigating hours of work accidents in the Australian coal-mining industry (where no clear pattern was found). This is largely due to the nature of operator fatigue but also partly due to the precise accident- and incident-reporting systems used.

Because of the difficulties in identifying whether fatigue was a primary cause of an accident, it is hard to fully estimate the percentage of accidents due to fatigue. In truck driving on public roads, figures of up to 20 percent of all fatal crashes involving fatigue as a main contributory factor are not uncommon. Accident reports and safety alerts in mining often mention fatigue as a factor; for example, a safety alert produced in New South Wales, Australia, following the fatal injury of an operator when changing a tyre mentioned the risk factor of fatigue (Department of Mineral Resources, 2004).

Similarly, fatigue produces decreases in performance and more near misses. Often operators are slower and/or more variable in their performance. This can be a significant issue when operating or maintaining complex mining equipment. Of course, there are other negative effects of fatigue such as more frequent health complaints and decreased work motivation.

10.1.3 Working hours in mining

The number of hours worked by many mining operators and maintainers has generally decreased over the last 100 years. With the increased pressures of decreased manning, fly-in fly-out operations, or long drives to and from mine sites at the start and end of each shift, many operators still work an excessive amount of total hours (although, of course, with large individual variations). However, there is nowadays a fairly widespread acceptance that long hours are a major factor in operator fatigue, and contribute to errors, incidents, and accidents at work or when driving to and from work (Mine Safety and Health Administration [MSHA], 2001).

Looking more at time on task, four hours working continuously on a single task that requires high levels of concentration (e.g., operating mobile equipment) can increase errors and incidents. Such errors might be "micro events," like not responding to a communication device. These are often more marked after seven to eight hours on shift (especially where an operator had an insufficient amount of preceding sleep). Therefore, as will be seen below, planning correct working periods and breaks is essential.

Interestingly, in the industrial environment, many studies have found that increasing the length of the working day does not increase output. For example, increasing the working hours from eight to ten for a heavy manual job may even reduce overall daily output, as workers tend to "pace" themselves in longer working days. But this is, of course, not so for machine-paced work (i.e., some mining tasks), where it is impossible to ease up, and instead greater fatigue may result.

Likewise, although the evidence is not conclusive (see Cliff and Horberry, 2008), shift durations of over eight hours often tend to increase the risk of errors and incidents (MSHA, 2001). However, although limiting shift length may be physiologically optimal, it needs to be balanced against practicalities: tasks still need to be performed (especially maintenance tasks), some mining equipment cannot easily be shut down, and pay is sometimes higher for longer shifts.

Also, in these days of increased automation and de-manning, operators may feel compelled to work longer hours. Regulating hours worked is theoretically possible, but prescriptive hours of work are frequently not ideal, and more flexible alternatives (such as fatigue management) are often preferable and are being introduced (e.g., by the New South Wales Mine Safety Advisory Council, Australia, in December 2009).

10.1.4 Nightwork

As we all know, nightwork and other shiftwork are virtually indispensable in the minerals industry; often it is economically essential to run a twenty-four-hour operation. Humans are naturally inclined to sleep at night (especially when it gets dark) and be active (work and play) during the day (when the sun rises). Many human bodily functions fluctuate in a twenty-four-hour cycle called the *circadian rhythm*; these include core body temperature and blood pressure. These bodily functions usually peak during daylight hours and are lowest at night and early morning hours. This trend is associated with not just physical functions but also alertness levels. Within this context, sleep is the most important influence on the circadian rhythm.

Most adults need 6–8 hours sleep per night for health and well-being. Often the daytime sleep of nightworkers is of poor quality (especially when taken in the noisy conditions that might be experienced at a camp at a mine site) and so has less recovery value. It is very difficult to alter circadian rhythms completely; often several weeks are needed to do this, so where they are needed, quick rotation shifts are now often recommended in the minerals industry.

Many studies have found that more work errors (leading to incidents or accidents) are likely to occur at night, with the 12:00 midnight–4:00 a.m. period especially prone to problems when an operator often needs to fight to stay awake. As much as possible, where dangerous or demanding tasks involving mining equipment operations or maintenance need to be performed (e.g., scheduled maintenance tasks involving multiple operators and a large amount of complex support equipment), they should be done when personnel are most likely to be alert (Cantwell, 1997). However, this of course can be difficult to achieve. Where fixed night schedules are likely, then adaptation to them is possible, especially if the operator is helped to resynchronise by means of meals, exercise, social interaction, and light. Such adaptation can help improve the operator's health and job satisfaction.

Nightwork can cause many health problems, such as stomach troubles, sleep problems, ulcers, and heart problems. Indeed, recent reports in the media have linked some types of nightwork to increased cancer rates. It is estimated that two-thirds of all night-shift workers experience some health complaints at some time due to their working hours (Kroemer and Grandjean, 1997). Night- and shiftwork can also cause social problems: less family time, little social contacts, and less opportunity for clubs and social groups. The world often seems made for day workers, so mining shiftworkers sometimes feel on the edge of society. Night shifts are often worse for older people because they are less able to cope and less able to recover from the physiological and psychological stresses that such shiftwork brings. Because of this, some operators, especially older ones, in the mining environment leave the industry, or transfer to day working.

Of the rest, many are "positive-choice" mining shiftworkers (i.e., they enjoy such work, and have fewer health complaints). Some positive aspects of shiftwork and nightwork can include less strict supervision, more autonomy, better pay, and sometimes more leisure time.

10.1.5 Strategies to combat operator fatigue

10.1.5.1 Naps and coffee

Where appropriate, two short-term strategies for helping to manage fatigue associated with operating or maintaining mining equipment are short naps and/or coffee. However, the caffeine in coffee, when used incorrectly to excess, can cause both a dependency and a tolerance to develop, and can also modify sleep patterns. Techniques such as fresh air, showering, communicating with colleagues, and stretching are not usually effective for more than a few minutes, and are often not possible in the mining environment. The best solution is to obtain proper (i.e., good-quality) sleep, but for night shiftworkers at a mine site this can be difficult due to daytime noise and light. Where it can be done, preventing fatigue from occurring in the first place (e.g., through better designed shift schedules, etc.) is, of course, the best way to control the risk, especially in a high-hazard industry like mining.

10.1.5.2 Fatigue management

One important issue is that once in a fatigued state, an individual operator finds it difficult to estimate his or her own abilities to perform a task. Most operators are aware of the early sign precursors of fatigue (e.g., yawning and stretching), and so ideally must stop work at that point, especially if the task is safety critical. Nobody is immune from fatigue, but sometimes the culture in the minerals industry is still for an operator to deny being fatigued until long after the symptoms have manifested themselves (for example, by dozing off or making frequent lapses). Also, in the real world, work cannot easily be stopped: it may not be practical (imagine the problems of stopping half way through a re-fuelling task), or may not even be permitted. With the increasing use of more "enlightened" fatigue management techniques, an operator is more encouraged to report that he or she is fatigued. Such fatigue management programs generally:

- inform workers, schedulers, and managers of the dangers of fatigue;
- encourage operators to stop if feeling sleepy, when possible;
- provide better information about sleep hygiene;
- require medicals (e.g., to detect sleep disorders); and
- require comprehensive records and audits.

10.1.5.3 Rest breaks

From the scientific literature (and from the personal experience of many mine workers), it is consistently found that rest breaks in shifts are very necessary—often they can actually increase daily output, despite slightly less hours being "worked" (as noted by Grech et al., 2008). There are different types of breaks and pauses, such as prescribed ones (e.g., a maintenance worker is relieved of duties for a twenty-minute lunch break), or more "implicit" breaks (e.g., an operator does something different, or a break due to the nature of the work such as a haul truck driver waiting for the truck to be loaded).

Kroemer and Grandjean (1997) stated that pauses or breaks can be especially good for the following:

- Training (so a new operator has time to reflect on techniques just learnt, a break of thirty minutes or more can be useful).
- Extreme machine-paced work. This is especially for the older workers, who often would ideally work slightly slower.
- Heavy work (so as not to exceed daily physiological capacity for work, e.g., in terms of energy expenditure, a couple of 15–30-minute breaks in an eight-hour shift is sometimes recommended).
- Social contacts (when a mining operator is working in "isolation").
- Heat (to allow the body to cool down).
- Close visual work (e.g., to prevent temporary myopia or eyestrain amongst maintenance workers undertaking a visual inspection of, for example, components for wear or corrosion).
- Mental work (which benefits from regular short breaks, even if only 5 minutes per hour).

From this, where shiftwork (especially nights) is needed on the minerals industry, a few recommendations (adapted from Kroemer and Grandjean, 1997) are as follows:

- Ensure adequate break(s) per shift for nourishment and short rest.
- Where legally possible, only use operators aged between twenty-five and fifty for shift- and nightwork.
- Only employ healthy, emotionally stable workers.
- Have at least a twenty-four-hour break after working a series of nights (especially if two or more nights are in the series), and plan some free weekends into the shift schedules.
- Avoid or limit extreme machine-paced work, where the operator has no flexibility in setting the pace, and fatigue results more quickly.

Of course, some of these recommendations are not practical in every situation, but they do provide some guidance about what is best practice in mining equipment operations and maintenance.

10.1.5.4 Fatigue detection technologies

Due to a greater awareness in the mining community of the danger of fatigue, the goal of reliable and sensitive fatigue detection technologies is very attractive. In general, fatigue detection technologies for operators (especially mobile equipment operators) are of four types (Hartley et al., 2000), although some systems now use a combination of these four measures (Wright et al., 2007):

1. Fitness for duty systems (which might measure an operator's state at the beginning of their shift)
2. Mathematical models of alertness (that can be used to evaluate if a roster is likely to cause excessive fatigue)
3. Systems that detect fatigue through changes in an operator's behaviour (for example, less steering wheel movements when driving a haul truck)
4. Systems that detect fatigue by monitoring the individual's state (for example, head nodding or eye blinks)

Similar to other occupational domains such as road transport or defence, a great deal of research and development is taking place in this whole area in the mining industry. However, it is being increasingly realized that it not enough just for a technology to be able to be able to detect fatigue; the following is equally important:

- That the device is unobtrusive (Wright et al., 2007). For example, it is usually not feasible or acceptable for an operator to have to wear a bulky fatigue detection device. Some devices to record an individual's state do so by means of special glasses worn by the operator. Although not ideal for all activities, the size of these glasses (similar to conventional eyeglasses) may make the technology acceptable in this regard.
- Likewise, the devices must be robust—capable of being used in harsh conditions such as underground, in bright sunlight, or during maintenance work.
- Finally, how the fatigue information will be used needs to be carefully considered (Hartley et al., 2000). For example, can the results of such technologies be used by mine management (e.g., for selection, training, or statistical purposes) or only the individual operator? Likewise, should the individual operator always stop work immediately if the device tells her she is fatigued? For example, if a truck driver was on a four-hour journey to deliver a load and the

device informed him in the last ten minutes of the journey that he was fatigued, would the operator stop immediately, or continue to finish the task?

So significant issues still exist. Such devices are attractive in many ways, but they should not be considered as a replacement of other fatigue management measures in mining.

10.2 *Mental workload*

Manual tasks are still carried out in most mines, for example in maintenance or when carrying smaller items of equipment. So, the physical workload is certainly still an issue in mining. However, the focus in this section is upon workload of a more cognitive and/or perceptual nature, which will be called *mental workload* here. The term *mental workload* is also often used when referring to longer term issues such as job burnout; however, it is the shorter term version that is of most concern in this book about mining equipment.

Like fatigue, mental workload is another major performance-shaping factor that impacts the safety and efficiency of mining tasks. Put simply, it can be defined as the amount of mental effort that is required by a mining operator in order to perform a task(s). As tasks (and equipment) become more complex, the required mental workload generally increases, or at least becomes more variable.

10.2.1 *Levels of mental workload*

Operator performance is generally best at intermediate levels of mental workload. If a task is too hard, obviously performance suffers; on the other hand, if a task is too easy, arousal and alertness may drop, and again performance will suffer.

Mental workload is an important issue to consider when assessing the demands placed on an operator by changes in a mining task. As reviewed earlier in the book when discussing new technologies, the task may change because the technology has changed (for example, communication devices are introduced), the environment has changed, or other tasks need to be carried out at the same time, such as speaking on a handheld radio whilst operating equipment.

In the minerals industry, mental workload can vary due to a large number of factors, which include the following:

- The time allocated to complete a task, where usually shorter periods of time allocated to a task lead to a higher workload. An example of this might be time allocated to load earth into a wagon.

- The number of tasks performed (sometimes simultaneously). An example here might be talking on the radio whilst monitoring displays in a mineral process control room.
- How difficult tasks are. For example, routine maintenance, compared to diagnosing a complex fault.
- The required level of accuracy and proficiency. For example, driving in an unrestricted space compared to driving in a narrow roadway.
- Individual factors: skill, intelligence, stress, and personality. For example, expert operators might experience a lower workload level than novices.
- Temporary personal factors such as fatigue, alcohol, or drug use, and minor illnesses such as colds and influenza—all of which generally increase workload.
- Environmental factors (e.g., heat or noise). As seen in Chapter 6, the additional strain caused by excessive heat or noise can have negative effects. Similarly, whole-body vibration from mobile mining equipment can influence perceived workload and also reduce reaction time (Newell and Mansfield, 2008).

10.2.2 *Mental workload as an interaction of person, task, environment, and equipment*

Mental workload can therefore be viewed as an interaction between the mining task, environment, equipment, and person. So, the important point to note is that it is not just the design of the equipment that increases or decreases mental workload, but the interaction of these individual, task, environmental, and equipment factors. That said, as seen in the discussion of new technologies earlier in the book, equipment design can have a major impact, especially when poorly designed or integrated.

In some tasks (such as routine maintenance), there are ways operators can reduce task load by, for example, slowing down; however, it is more difficult to reduce mental workload for many tasks (e.g., those machine or team paced).

10.2.3 *How to measure it?*

The scientific literature states that there are three main types of workload measures. These vary in practicality for mine sites or equipment designers to use:

1. Self-reports. At its simplest, this would consist of asking an operator how difficult or demanding her task is. More sophisticated and validated questionnaires (such as the NASA—Task Load Index) also exist and are commonly used.

2. Physiological measures (e.g., heart rate variability). Although often scientifically rigorous and valid, they are difficult to conduct at mine sites or similar locations (unless specialist equipment and personnel are present).
3. Task performance measures. This covers both primary task and performance on a secondary task (e.g., simultaneously holding a conversation whilst operating machinery). Examples here might include fuel efficiency or amount mined per hour.

10.2.4 Mental workload and new technology

Returning to the issue of new technology in mining, controlling workload is a key factor—too high a workload level can lead to the demands exceeding an operator's capacity to cope, whereas low levels can lead to operator boredom and being "out of the control loop." In many mining operations, often a problem is that the workload is unevenly distributed, rather than simply reduced: for example, at key times (such as responding to a series of alarms following a component failure) workload is excessively high, whereas on some occasions the technology reduces operators to passive monitoring of the system. This issue seems set to continue with increased mining automation, and careful consideration of the effects of the technology (and the new tasks required by that technology) upon the workload levels of operators is a key factor.

10.3 Occupational stress

10.3.1 Is a little stress a good thing?

It is often thought that a little stress can sometimes be a good thing; it can motivate an individual, and help them focus on a task or the equipment they are using. However, it is argued here that psychological stress, when appropriately defined, is always a negative factor. Like fatigue and mental workload, stress can therefore be another important performance-shaping factor when operating or maintaining mining equipment.

10.3.2 Effects of stress

Psychological stress means that operators may perceive tasks, and environmental and/or internal demands, as nearing the limit of or exceeding their resources for managing the situation.

In terms of its effects on task performance, stress can make people focus more narrowly on a few specific aspects of their task and neglect other aspects, possibly leading to negative consequences: for example, focusing on only one display in a process control room, or only looking out for vehicles straight ahead when driving a haul truck. Likewise, in the United States, Kowalski-Trakofler et al. (2003) found stress to be a major factor in individual judgment and decision making in underground emergencies, often by narrowing the individual's focus of attention.

Also, stress can increase the likelihood that a person may engage in unsafe, risky behaviours by adopting "shortcut" work methods, whereby safety rules and procedures are not properly followed. Examples might include speeding, not wearing personal protective equipment, or not correctly isolating a piece of equipment.

The implication is that if there are high time pressure and high stress, operators may think that taking risks is simply part of their job and that there is not always time to follow safe procedures. Shorter term reactions to stress may include "flight or fight" (e.g., abandoning the task or confronting colleagues), increased anxiety and agitation, more arguments between operators, and increased use of alcohol or illegal drugs. In the longer term, issues such as job burnout and increased health complaints can be evident (such as digestive or cardiovascular problems) (Grech et al., 2008).

10.3.3 Who are affected most by stress, and what helps?

As for mental workload, stress is often a particular issue for younger mining operators who are still learning many tasks, and the operation of equipment. It can also be a problem for older operators, who may have less ability to cope with task demands due to declining physical capabilities (e.g., strength or resistance to fatigue) and perceptual and cognitive capabilities (e.g., failing vision or impaired memory).

Having some degree of "mastery" (or control) over a situation helps reduce stress. Similarly, social support from family and friends is important for coping (Riordan et al., 1991). Remote mine sites are not always conducive to coping with stress unless the mining company provides additional support of some form, such as counseling, social events, or flexible work hours.

As stress is an individual phenomenon (which depends on how an operator perceives the demands and perceives her ability to cope with such demands), it is sadly not easy to provide many general task recommendations to reduce its occurrence. A universal "level" of stress for all operators at a mine site does not exist. Further, stressors like being away from one's family and friends during long periods of work (e.g., for mine workers in remote locations) are difficult to manage. The support networks in these situations can often be improved by greater communication between operators and their supervisors or medical officers to help identify risks. This might be instigated as part of formal safety management programs which focus on aspects such as open communication.

10.3.4 Stress measurement

Due to the "psychological" aspect of stress—whereby it is not a set level that is the same for everybody—stress is often best measured by individual questionnaires. The influential UK stress researcher Tom Cox has produced several of these. Another influential version by Lemyre and Tessier (2003) is shown in Table 10.1.

10.4 Distraction

The final performance-shaping factor that will be considered in this chapter is distraction. Unlike fatigue, until recently there was relatively little work done examining the effects of distraction upon occupational work performance. That is beginning to change; however, most of the work done on distraction has been related to driver distraction in the road safety domain. Clearly there are many parallels between driver distraction from driving a truck on the public roads, and operator distraction from driving a truck at a mine site. As such, driver distraction, and its links to mining equipment (particularly mobile equipment), will be considered below.

10.4.1 The importance of driver distraction

Driver distraction is now becoming an important human factors research area in the United States, Continental Europe, the United Kingdom, and Australia. This is not surprising when the issue was implicated in half of the 25 percent of crashes involving driver inattention in the United States (Stutts et al., 2001). For example, at the time of writing this chapter:

- A major book was published about the issue of driver distraction in 2008 (Regan et al., 2008).
- International conferences on the issue recently took place in Australian, Canada, and Sweden.
- A major driver distraction workshop took place in England, sponsored by the UK road authorities.
- The US National Highway Traffic Safety Administration has identified driver distraction as a high-priority area.
- The European Union funded several major research projects considering different aspects of driver distraction (e.g., due to new technologies being introduced into vehicles).

Table 10.1 Psychological Stress Measure (PSM-9)

Mark the number that best indicates the degree to which each statement has applied to you recently, that is, in the last 4–5 days.

Description of mood	Not at all	Not really	Very little	A bit	Some-what	Quite a bit	Very much	Extremely
	1	**2**	**3**	**4**	**5**	**6**	**7**	**8**
I feel calm.								
I feel rushed; I do not seem to have enough time.								
I have physical aches and pains: sore back, headache, stiff neck, and stomach ache.								
I feel preoccupied, tormented, or worried.								
I feel confused; my thoughts are muddled; I lack concentration; I cannot focus.								
I feel full of energy and keen.								
I feel a great weight on my shoulders.								
I have difficulty controlling my reactions, emotions, moods, or gestures.								
I feel stressed.								

Source: Adapted from Lemyre and Tessier (2003) and Lemyre et al. (1990).

10.4.2 *Definitions of driver distraction*

Although no formal definitions of *driver distraction* have been widely agreed upon (Horberry et al., 2008), most researchers in the field would accept that distraction occurs when a triggering event induces an attentional shift away from driving (Stutts et al., 2001).

As such, the two key elements are as follows:

- Attentional shift caused by a
- triggering event.

Similarly, Regan et al. (2008) stated, "Distraction is the diversion of attention away from activities critical for safe driving towards a competing activity" (p. 624).

This definition would also include "cognitive" distraction where the operator is thinking or worrying about something. Using these definitions, driver distraction is also a potential problem regarding operating mobile equipment at mine sites. Further, without too much imagination, it is possible to see how distraction might be an issue when operating or maintaining almost any mining equipment. For example, a maintainer changing an oil filter is distracted by two colleagues having a nearby conversation, or an underground operator is distracted by money worries at home.

10.4.3 *Internal or external distraction*

Returning to driver distraction, attention can of course be captured by events within or outside the vehicle.

- Within-vehicle distractors include many of the devices that car, truck, and professional drivers now use, such as phones, radios, stereos, and navigation systems, as well as passengers and other equipment. Clearly, this is also an issue for mobile mining equipment, often due to the environment in which the equipment is operated. An example of this is a significant incident that occurred in Queensland, Australia, where a dozer ran over a light vehicle; one of the reasons given for this was that the dozer operator did not hear a radio notification due to the noise level within his cabin that necessitated him wearing hearing protection—in this case, the distractor was the wearing of hearing protection (Queensland Government Mines Inspectorate, Safety and Health Division, 2001).
- Outside vehicle distractors in the road environment can include traffic signs, other vehicles, pedestrians, and billboards. Compared to the road environment, this might be less of an issue in mining (especially when underground), but still a possible factor.

10.4.4 Distraction minimisation

The number of possible distractions in many mining environments is therefore fairly large, so producing a definitive list to help minimise them is difficult. However, some aspects to consider are as follows:

- When new equipment is introduced (e.g., new communication devices in underground mining), how can they be integrated with existing tasks, and how do they alter existing tasks?
- The design of the operator's job, for example regularly changing the tasks being performed. This can help prevent underload (and the tendency for an operator to become distracted by talking to colleagues) or "cognitive distraction" (e.g., thinking about the spouse at home).
- Safely designing equipment to minimise distraction; for example, by moving low-level warning indicators away from high-level critical warnings.

10.5 Conclusion

This chapter has focused on wider organisational and task issues related to mining equipment. In particular, it examined four possible performance-shaping factors: first it considered fatigue and shiftwork, and then it more briefly reviewed mental workload, and stress and distraction. It should be noted that other organisational and task factors can also have an impact upon mining equipment safety and performance. Examples of these include work organisation (e.g., supervisory style) or task-linked factors (such as frustration, or unclear task goals); however, these were not reviewed here due to them being less closely linked to mining equipment.

Chapter 11 builds on this and considers other factors that are linked to mining equipment safety and efficiency, in particular considering the role of training.

chapter eleven

Training

Co-written with Jennifer G. Tichon
The University of Queensland, Australia

11.1 Why train?

The main aims of considering human factors in the design of mining equipment are to ensure that equipment is as easy to operate and maintain as possible, and that human-related risks when interacting with the equipment are designed out as far as possible. However, regardless of how well equipment is designed, there will always be a need to provide some degree of training, in both the equipment use and maintenance. This need is particularly acute for new or automated mining equipment (as described in Chapter 9), where such technologies may require quite different operator skills and ways of working compared to previously.

Training provides the skills required to operate and maintain the equipment in the manner intended by both the manufacturer and the mine management (site and corporate). With practise (and feedback), skills become automatic, and this automaticity frees the operator to attend to sensory information indicating the development of abnormal situations, potentially in time to alter behaviour, or take action in time to avoid unwanted consequences, be they injury, equipment damage, and/or production delays. Training also provides the knowledge and cognitive skills required to undertake problem solving such as fault diagnosis, or choosing the correct response in an emergency situation. Again, practise and feedback are required.

11.2 Human factors in the design of training

Instructional system design models (e.g., Gordon, 1994) exemplify the application of human factors principles to training, and include similar features to the human factors equipment design process (mentioned in Chapter 2 of this book). In essence, such models involve front-end analysis steps (analysis of the situation, task, equipment interface, trainees, training needs, and resources, leading to definition of the training functional specifications), followed by design and development steps (training concept generation, training system development and prototyping, and usability testing) and system evaluation steps (determining training evaluation criteria, collection and analysis of these data, and subsequent modification of the training if indicated).

The front-end analysis (or training needs analysis) step in training design is critical. In particular, a comprehensive analysis of the tasks performed by equipment operators and maintainers is required before the training needs and associated functional specifications can be determined. The aim of the task analysis is to describe the knowledge, skills, and behaviours required for successful task performance, and identify the potential sources and consequences of human error. This task analysis would typically involve interviews with experts, reviews of written operating and maintenance procedures, and observations of equipment in use. It should include consideration of the information required by equipment operators and maintainers and how this information is obtained, the decision-making and problem-solving steps involved, the action sequences, and attentional requirements of the task. The task analysis should be conducted systematically, and well documented, to provide a solid foundation for the design of training and to provide a template for future training needs analyses.

On its own, training is never an adequate control for foreseeable human errors which have potentially serious consequences (Simpson et al., 2009). Where such errors are identified during the task analysis, this information should be fed back into the equipment design process and effort should be expended in identifying design control measures (e.g., by using the operability and maintainability analysis technique [OMAT], described in Chapter 2). In this way, a systematic task analysis can be seen to have dual functions, and be an integral part of both equipment design and training design.

An extension of the task analysis to include a "cognitive" task analysis may be justified for more complex task–equipment interfaces. Cognitive task analysis seeks to understand the cognitive processing and requirements of task performance, typically through the use of verbal protocols and structured interviews with experts. The outcomes of a cognitive task analysis include identification of the information used during complex decision making, as well as the nature of the decision making. The cognitive task analysis can also reveal information which will underpin the design of training and assessment. Again, the outcome of a cognitive task analysis may include identification of design deficiencies which should be fed back into the design process.

The results of the task analysis are also used in the second phase of training design to define the actual contents of the training program, as well as the instructional strategy required. Regardless of the content of the training (the competencies required) or the methods employed (e.g., simulation), most effective instructional strategies embody four basic principles:

1. The presentation of the concepts to be learned
2. Demonstration of the knowledge, skills, and behaviours required
3. Opportunities to practise
4. Feedback during and after practise (Salas and Cannon-Bowers, 2001)

An initial training design concept is typically refined iteratively through usability evaluation of prototype training models, until a fully functional final prototype is considered ready for full-scale development. Issues to be considered include the introduction of variation and the nature and scheduling of feedback. A compelling case has been presented (Schmidt and Bjork, 1992) to suggest that variation in the way tasks are ordered and in the versions of the tasks to be practised is important, and that less frequent feedback should be provided. Whilst immediate performance may be reduced, retention and generalisation are enhanced as a consequence of the deeper information processing required during practise. For example, Swadling and Dudley (2001) describe the use of a virtual simulation to train operators prior to the introduction of remotely controlled load-haul-dump (LHD) vehicle in an underground metalliferous mine. Human factors training research would suggest that such training should ideally include increasing variability (including both normal operation, and abnormal situations such as virtual equipment malfunction), and periods of practise in which knowledge of performance feedback was withheld.

Evaluation of the consequences of training is also an essential and non-trivial step, and the task analysis aids in determining the appropriate performance measures to be used in evaluation (or competency assessment). A valid training evaluation requires careful selection of evaluation criteria and measures (closely connected to the task analysis results), and systematic collection and analysis of data. This, in turn, might suggest improvements to the training process for subsequent occasions—so the whole process often involves a degree of iteration.

11.3 Expertise and training

Expertise is not easily acquired in any sphere—certainly this is true regarding operating and maintaining modern mining equipment. Attaining expert performance across a variety of domains has been estimated to typically require about ten years, 10,000 hours, or millions of trials (Ericsson et al., 1993). Task analysis typically involves studying experts; however, studying the differences between experts and novices, and the development of expertise, has also been proposed as a means of deriving principles for training design (e.g., Abernethy, 2001).

Comparing the behaviours of novices and experts can assist in identifying the factors which limit the performance of novices. Identification of the factors which do, and which do not, discriminate the performance of experts and novices provides guidance towards the aspects of the task towards which attention should particularly be directed during training. Comparison of the information sources used by novices and experts, for example, can help determine what information must be learnt by novices to become an expert in a given task.

11.3.1 *Sensation and perception differences*

Across a range of skills and domains, there are some consistent findings relevant to the design of training which emerge from the study of expert–novice differences, expertise, and skill development. For example, it would be unsurprising to note that persons with poor sensory reception may have difficulty performing mining tasks which require the acquisition of information via the sense in which they have a specific deficit. People with visual defects perform poorly on inspection tasks requiring high visual resolution, for example, and this suggests an obvious need for occupational screening. Any screening must be very specific to the perceptual abilities required during the task, however. For example, assessment of static visual acuity is inappropriate for tasks that require assimilation of dynamic visual information, such as driving mobile mining equipment.

Whilst poor perceptual ability may limit performance on some equipment-related mining tasks, the corollary is not true—there is no evidence that experts are characterised by above-average sensory abilities. Attempts to enhance sensory abilities through training are misguided. Where experts and novices do differ in sensory terms is in the ability to make sense of the sensory information received, that is, in perception. Task-specific measures of perception across a range of domains (mining included) indicate that experts are superior at discriminating perceptual events such as flaws (Blignaut, 1979), better able to recognise patterns, and better able to predict future events based on current sensory information. In mining, this is particularly relevant to the ability to perceive the probability of hazards, such as roof fall, on the basis of vision of the roof and rib conditions. Expert mining equipment operators might similarly be expected to make use of engine or other noises as indications of the state of the equipment, or the likelihood of future events such as a stall or overheat. Other cues such as the colour of water coming from a drill hole provide the expert bolter with information about the strata being drilled through.

Some evidence exists to suggest that acquisition of these perceptual skills can be accelerated through training (Starkes and Lindley, 1994; Williams and Grant, 1999), particularly through the use of video-based training or virtual reality (VR) simulations. However, a potential barrier to designing training to achieve this is that the perceptual expertise appears to be largely implicit, that is, not accessible to the consciousness of the expert, not able to be verbalised, and therefore difficult to define. This is particularly the case in operating and conducting routine maintenance of mining equipment, where the skills are often "over-learnt" and with a high degree of automaticity by experts. For example, verbalising how to drive a loaded haul truck up an incline can be difficult for an experienced operator to do, except in the most general terms.

11.3.2 Decision-making differences

Another area in which experts and novices differ is in decision making. Experts are typically faster at problem solving (Crandall et al., 2006), have better short- and long-term memory of relevant states or events, and have more knowledge of relevant facts and procedures. Experts are also more likely to see and represent problems (such as fault diagnosis) at a deeper, more first-principles level, and spend greater time analysing problems before arriving at a diagnosis of solution (Gilhooley and Green, 1988; Glaser and Chi, 1988; Ye and Salvendy, 1996). Whilst it is attractive to try to facilitate the acquisition of such cognitive skills, such as fault diagnosis in an equipment maintenance context, by providing novices with access to experts' procedural knowledge structures, evidence for success in this is scant.

11.3.3 Action differences

Differences between experts and novices in reaction and movement times are evident for tasks which require rapid responses, although these differences may reflect greater perceptual skill leading to more rapid assessment of the response required. No differences are found in simple reaction times, and experts are constrained in the same way as novices where rapid serial responses are required. In terms of movement production, both the movements and forces produced by experts are less variable than those by novices, and experts harness the complex intersegmental dynamics of the body and produce more efficient, and less effortful, body movements.

11.3.4 Attention differences

As mentioned at the outset of this chapter, one of the characteristics of skill development is increasing automaticity. The consequence is a freeing up of attentional capacity to deal with additional information. For example, the novice driver's attentional capacity is all but exhausted by the management of the essential vehicle controls (brake, gears, and accelerator) and the primary goal of maintaining course through the environment (especially in confined underground environments, for example), and little or no attentional capacity remains for watching ahead for potentially hazardous situations. Conversely, with practise, these essential vehicle control functions become automatic, and proficient drivers are able to attend to the road ahead, identifying hazards and taking necessary actions well before arriving at the hazard.

11.4 *Issues associated with training*

Mining equipment–related training can be delivered through many different modes, from written materials, audiovisual materials of varying complexity, and computer-mediated training of varying degrees of interactivity, through simulations of varying fidelity, to full-blown 360° stereoscopic virtual reality simulations incorporating tracking of the trainees' movements, and an infinite variety of combinations and permutations of all of the above.

There is some evidence that a conventional approach to training of presenting knowledge and theory before interactive training (whether simulated or "on the job") is not the most effective because teaching via lecture or text involves learning declarative knowledge in isolation from its function in task performance with the mining equipment. A more effective method of training tasks with relatively high cognitive complexity is to intersperse presentations of small amounts of declarative knowledge with a series of graduated simulations (although retaining relevant perceptual cues). As training progresses, the simulations become more complex (Mumaw and Roth, 1995). This approach would seem to be applicable for the training of many operational and maintenance tasks in mining.

11.5 *Use of simulation in training*

A review of traditional training methods used in mining (Churchill and Snowden, 1996, cited by Schofield et al., 2001) suggested a number of potential problems, including the following:

> [R]ote learning of information is the most common technique used by trainers with the same sets of training media being used from year to year. Many teaching methods present too much material, too rapidly, with little or no opportunity for worker involvement.
>
> Trainees frequently fail to attend to the problem at hand, often dividing their attention between what is going on at the front of the classroom and interpersonal interactions with those around them. …

> Skill degradation is an important issue. When the hazards of a mine environment are combined with the issue of skill degradation, the need for realistic training becomes paramount. (p. 154)

Schofield et al. (2001) proposed that virtual reality simulation offered the opportunity to improve safety-related training in mining, suggesting that "the capacity to remember safety information from a three-dimensional computer world is far greater than the ability to translate information from a printed page" (p. 155).

There is no doubt that virtual reality simulation offers the opportunity to develop both perceptuo-motor skills and cognitive skills such as problem solving, decision making, and hazard perception, without exposing trainees or others to unacceptable risks (a number of fatalities have occurred in mining during on-the-job training). This strategy has been employed in other hazardous industries such as aviation, rail, health, and defence. The simulators range in cost, fidelity, and functionality. Simulations are ideal to train tasks which have large perceptual or pattern recognition components, tasks which require teamwork or group problem solving, and tasks which are too dangerous or expensive to undertake for training alone. The potential for improved safety suggested by Schofield et al. (2001) and others (e.g., Bise, 1997; Filigenzi et al., 2000; Wilkes, 2001) has been embraced by the mining industry, and virtual reality simulation is beginning to be adopted. Kizil (2003), for example, suggests, "There is no doubt that the use of VR based training will reduce these injuries and fatality numbers" (p. 569).

Virtual environments have been used as a training medium for many years in aviation, medicine (and particularly surgery), and defence. Considerable evidence exists to demonstrate the effectiveness of virtual reality for pilot training (see Carretta and Dunlap, 1998, for a review). Whilst flight simulators have been consistently demonstrated to result in skill acquisition by pilots, the effectiveness of the training is strongly influenced by the task to be trained and the amount and type of training provided. Simulators have been found to be more effective in training for take-off, approach, and landing than for other flying tasks. Landing skills, and instrument flying learnt in a simulator, have also been shown to transfer to the real task (Pfeiffer et al., 1991; Hays et al., 1992). Flight simulators are typically used to complement flying time rather than substitute for flight time; however, there is evidence that simulators reliably produce superior training compared to aircraft-only training (Jacobs et al., 1990).

Strong evidence also exists for the use of virtual environments for training a variety of surgical skills (see Gurusamy et al., 2008, for a review) and the use of simulators for surgical training is becoming a standard feature of medical education. Similarly, evidence exists which demonstrates the benefits of virtual reality simulations in training for drivers of cars (e.g., Turpin and

Welles, 2006), trucks (Parkes and Reed, 2006), snow plows (Kihl and Wolf, 2007), trains (Tichon and Wallis, 2009), and emergency vehicles (Lindsey, 2005) in terms of both safety-related behaviour and fuel efficiency.

In addition to training miners to operate equipment in normal conditions, virtual environments also offer the potential to train miners to cope with emergency situations. Enhancing such emergency response skills has been considered important by a number of industries. Of particular interest are those cognitive skills known to degrade under stress such as critical thinking and decision making.

The development and evaluation of virtual environments for training in decision making under stress have a strong empirical basis provided by the military. In 1998 the US Office of Naval Research completed a seven-year research project focusing on decision making under stress (TADMUS) (Cannon-Bowers and Salas, 1998). Rather than focussing on skills development, the focus here is on ensuring that performance does not deteriorate under stressful conditions, hence stress exposure training (SET) (Driskell and Johnston, 1998).

Such training focuses on developing those cognitive skills required to maintain effective performance under degraded work conditions. The aim is to incite the same emotional reaction and stems from research demonstrating that for some tasks normal training procedures did not improve task performance when the task was later performed under stressful conditions (Zakay and Wooler, 1984; Tichon and Wallis, 2009).

Negative, affective states such as stress can impair decision making by causing an overestimation of risks which results in unprofessional or unsafe choices based on perceived levels of danger to self and others—errors which Peters et al. (2006) termed "avoidant choices." Sub-goals of stress training include gaining specific knowledge of, and familiarity with, the operational environment to assist trainees to form accurate expectations. This will improve trainees' ability to predict outcomes and avoid errors, as well as decrease their propensity to be distracted by novel sensations (Driskell and Johnston, 1998). The overall goal of cognitive training via virtual reality is to build confidence in trainees in their ability to perform under adverse conditions.

11.6 VR simulation training in mining

Virtual reality has been identified as a potential avenue for training in the mining industry for some time (Bise 1997; Filigenzi et al., 2000), and was identified as a desirable research focus by the US National Research Council, Committee on Technologies for the Mining Industry (2002).

As noted earlier, Swadling and Dudley (2001) described an application in which operators' performance whilst driving a virtual simulation of a remote LHD operation (VRLoader) was compared with the operators' subsequent performance during driving the remote LHD. The simulation was found to be an effective training tool, and performance on the simulation was predictive of subsequent performance whilst driving the remote LHD. A range of equipment simulators including dozers, draglines, haul trucks, shovels, continuous miners, longwalls, and roof bolters are available from commercial vendors, with others under development.

A jackleg drill simulation (MinerSIM) aimed at training new operators (Dezelic et al., 2005; Hall et al., 2008; Nutakor, 2008) has been constructed. MinerSIM consists of a web tutorial and a virtual reality simulation which allows trainees to install rock bolts in a virtual environment. The simulation provides exposure to both normal and abnormal situations. Based on the results of evaluations of equipment operation in other domains, it is likely that equipment simulators will be effective in assisting trainees to develop the perceptuo-motor expertise required to operate the equipment, and that this will reduce the real-world practise necessary to achieve competent operation. This has potential safety benefits both for the trainee and for others located in the vicinity of the equipment.

A virtual conveyor belt safety-training program has also been described (Lucas et al., 2008; Lucas and Thabet, 2008). The simulation consists of an instructional module, and a task-based training module in which the trainee completes assigned tasks. Both desktop and immersive versions have been described. A similar application was described by McMahan et al. (2008) in which training in pre-shift inspection for haul trucks was provided in both desktop and immersive virtual environments. The training included a "virtual tour" which introduced the information necessary to conduct a pre-shift inspection (parts to be inspected and explaining defects). The trainees then completed a virtual inspection, and were shown a simulation of the potential consequences of overlooking defects.

The ability to detect and identify hazards has been another target for training in virtual environments (e.g., Filigenzi et al., 2000; Orr et al., 2008). Squelch (1997) provided hazard awareness training via desktop virtual reality. A comparison with traditional training methods was attempted for two groups of thirty miners. Whilst the trainees reported preferring the virtual reality training, no quantitative comparison between the two training media was possible. Denby et al. (1998) similarly trained mine operators in hazard identification and hazard avoidance using a desktop virtual haul truck, processing plant walkthrough, and underground fire and explosion; however, no evaluation was reported other than trainee reactions.

There is no doubt that virtual reality simulations have other potential roles in the minerals industry beyond training. Other uses include data visualisation, and improving equipment design through the exploration of virtual equipment models to assess safety-related design issues such as visibility and data visualisation (e.g., Delabbio et al., 2003), or consequences of design characteristics such as directional control–response compatibility (Zupanc et al., 2007).

In summary, a range of VR technologies have been developed and applied across a variety of equipment-related applications in the mining industry. Operator and maintainer training programs include general safety training for hazard recognition and escape planning in underground mining, educational programs based on accident re-creation for prevention, equipment operations training, and safety design reviews for new equipment and processing layouts. An unparalleled advantage of utilising VR in miner training is the lack of physical restrictions, which allow the trainee to be placed in a position where it is physically impossible to be, such as inside a piece of equipment or cut-away views of rock falls. Previously unseen points of view of incidents and accidents provide invaluable learning experiences (Schofield, 2008).

Having said that, exactly why simulation works is not well understood (Salas and Cannon-Bowers, 2001). Most evaluations which have been undertaken rely on subjective responses of trainees, rather than performance data, and more systematic evaluations are required. Simulation training should be developed with training objectives (derived from task analysis—e.g., Zupanc et al., 2009) in mind, and allow for the measurement of performance outcomes (Salas and Cannon-Bowers, 2001).

The following principles for obtaining maximum benefit from the use of virtual reality simulation as a training medium may be derived from current evidence:

- The use of simulation should be an integrated part of a training plan (derived from a systematic training needs analysis) that includes clearly stated goals, quantitative performance measures, and trainee feedback.
- Trainee feedback should be referenced to operational requirements.
- An event-based approach should be used in which trainees are presented with discrete scenario-based training events which allow practise of the specific skills (identified through systematic task analysis) that would be required in real-life situations.
- Trainees should be given the opportunity to make, detect, and correct errors without adverse consequences.
- Trainees should experience success or a sense of mastery during the training.
- Interaction with the simulation is a factor which contributes to high levels of presence and immersion, which in turn support the transfer of knowledge to the real world. Consequently, simulators should provide an experience in which trainees are active participants rather than passive bystanders.

- Skill development requires practise. Trainees should have opportunities for repeated practise under operational conditions similar to those in the real world.
- Simulator sickness may cause trainees to disengage from the immersion, and should be avoided as far as possible.
- Effective transfer to high-stress operational environments occurs in conjunction with higher levels of perceived affective intensity within the virtual environment, and consequently, the virtual world should aim to re-create the same stress levels.
- Fidelity may be necessary to ensure trainee acceptance; however, it is not necessarily required for effective cognitive skills training.
- Mixed fidelity approaches to the use of simulation are those in which combinations of different simulation technologies of varying degrees of immersion, coupled with practice in real-world environments, may provide the most effective training regime.

11.7 Conclusion

Training for the operators and maintainers of mining equipment is an integral part of ensuring the safe and productive use of the equipment. The design of training should encompass a structured process incorporating a training needs analysis leading to the definition of the functional specifications, an iterative design component incorporating usability testing, and evaluation. Training should include practise in a variety of circumstances and situations. The use of virtual reality simulation is a promising method for allowing the safe exposure of trainees to unusual, and otherwise high-risk, scenarios.

chapter twelve

Conclusions

12.1 Summary

This book has demonstrated the richness, issues, problems, and benefits of human interaction with mining equipment. Further, it is anticipated that the book contains practical information to help improve the safety and productivity of equipment-related mining operations and maintenance. Some of the key aspects covered included the following:

- The equipment design process, safety in design, and human element considerations in the equipment design life cycle
- Why it is vital to systematically consider actual operations and maintenance tasks at a site level
- The importance and problems associated with manual tasks
- Why older mine workers need to be considered
- Safety versus production trade-offs, and the current demands for bigger and more efficient mining equipment
- Equipment access and egress considerations and issues
- The importance of vision and the physical environment
- The design of equipment controls and displays
- Human interaction with new technologies and automation
- Organisational and training issues with respect to mining equipment

It has been shown that mining equipment should be viewed as one part in the wider system of work that also involves individuals, groups (work teams), the organisational environment, the physical environment, prescribed tasks versus how work is actually done, communications, the wider society and culture, as well as the mining equipment and technology being used (Grech et al., 2008). Whatever the different elements of the system, it should be clear that operating and maintaining mining equipment are done within a wider context of work, and that improvements and new equipment designs will be most effective if considered within this wider work framework rather than as piecemeal adjustments, retrofits, or add-ons.

12.2 *Future general trends in mining human factors*

Throughout this book, some future trends in mining human factors have been hinted at. To make this more explicit, it is anticipated that future important issues in this area will include the following:

- Refining human factors methods and tools: for example, prescriptive versus risk-based methods to help design, operate, maintain, and audit mobile equipment used in surface mining. It is hoped that the information contained in this book can be of considerable use to help guide this process. An example of this, concerning the use of human factors information for equipment procurement, is given in the next section of this chapter.
- Further consideration and integration of system-wide issues, including changing the mind-set to generally consider human error as a consequence, not cause, of mining equipment-related incidents. Moving away from the "train and blame" view that is still often present today will help to generate a better understanding of the human element in incidents, and so will ultimately help to develop better management systems, equipment designs, and other countermeasures to reduce the number and severity of such incidents.
- The twin aims of mining human factors (safety and health, and productivity and efficiency) will always be the key issues. However, it is important to sometimes consider these more broadly and over a longer time frame. For example, it is important to consider issues around motivational and emotional aspects of a task or interacting with a piece of equipment as these can have a strong influence on if the equipment is used correctly (and not misused, sabotaged, or ignored); similarly, it can lead to better work motivation (so often resulting in better productivity). More generally, using human-related qualitative data to a greater extent in mining is important.
- Developing and applying better cost–benefit analysis models related to human element considerations, and "selling" the benefits of human factors more. Again, it is anticipated that the information contained in this book can be of substantial assistance here.
- Similarly, a greater awareness by managers, supervisors, designers, dealers, contractors, regulators, operators, maintainers, and the general public of the benefits of applying a user-centred approach in mining human factors is a key goal. It is hoped that this will lead to an increased professional recognition in the minerals industry of the status of a "human factors engineer," "ergonomist," or "user-centred designer." This would in no way overshadow the traditional technical

disciplines of mining (e.g., geology, or mechanical engineering) but should complement them in the development of safer and more efficient mines of the future.

12.3 Future human-related trends in mining equipment design, operation, and maintenance

In addition to these above-mentioned trends in mining human factors, it is believed that the following three specific mining equipment–related issues will become of greater importance in the near future.

12.3.1 The need for better human factors design and procurement tools

Assuming the importance of human factors is further recognised by the mining industry as a whole, then human factors methods, know-how, and findings need to be used to a greater extent in the design, purchasing, set-up, and management of equipment. Taking one of these as an example, human factors in equipment procurement, then one way to achieve this is through functional specifications devised to assist in new equipment purchasing.

As the demands of the mining industry increase, technology must adapt to optimise efficiency and enhance productivity without compromising safety. Mine sites cannot keep purchasing the same equipment year after year. The OMAT process described earlier in this book is one tool that could be used to identify the human factors risks in mining equipment, but the primary focus for this technique was for equipment designers and manufacturers, and secondarily for mine sites to audit their existing equipment. The purchasing of new equipment could certainly benefit from additional tools, potentially combining risk-based methods and human factors best practice.

One way to develop such procurement tools is to build on current industry initiatives. For example, the Earth Moving Equipment Safety Round Table (EMESRT) Design Philosophies could be used to help develop functional specifications for operator mobile mining equipment, supplemented with other best human factors principles (such as those contained in this book). The resultant functional specifications developed then may be easily included in a mine site's tender documentation for mobile equipment procurement to ensure the delivery of a safe and efficient design from the equipment manufacturers. Ideally, these functional specifications would give broad best practice human factors information that would be applicable across a range of current- and future-generation mobile equipment.

The outcomes of such an approach would benefit the mine sites and their personnel through improved purchasing and eventually design of mobile mining equipment. More specifically, through original equipment manufacturers (OEMs) needing to address the issues contained in the procurement documentation, it would help to create best practice designs mobile equipment. Likewise, such a procurement tool could help to better control the purchasing and integration of add-on or retrofitted technologies (e.g., to reduce overload and distractions to operators from piecemeal add-on technologies). Such a process could also allow the functional specifications and procurement documentation to be a means for mine sites to explicitly and consistently communicate their requirements better to OEMs. Finally, and ultimately, significant financial benefits might result for a mine site and the industry as a whole in terms of less equipment-related injuries or workers compensation claims, less equipment repair costs and downtime, and an increase in production hours through improved designs for maintenance. Good human factors information is now available, but it needs to be used and disseminated correctly.

12.3.2 Error-tolerant equipment

A recent book by Simpson et al., 2009 investigated human error in mine safety. They looked at a variety of levels of what they called *predisposing factors* that make errors more (or less) likely. These factors were as follows:

- The person–machine interface
- The workplace environment
- Codes, rules, and procedures
- Training and competence
- Supervision and line management roles and responsibilities
- Safety management systems and organisational and safety culture

One of Simpson et al.'s main points was that although there might be some occasions where a front line operator or maintainer might be responsible for an error, the vast majority of errors are predisposed or even directly caused by "latent" factors that are either elsewhere in the organisation or from another source (such as equipment manufacturers, through less than adequate designs). They recommended a three-pronged approach that involved risk assessment for human error potential (for example, in new equipment), auditing and assessment of current operations and equipment, and systematic consideration of human factors in accident investigations (including equipment-related incidents). Ultimately they declared, "Human error is and always will be inevitable. However, to accept that its consequences are always an inevitability would be both foolish and dangerous" (Simpson et al., 2009, p. 137).

So human error in some form is therefore inevitable, but equipment can still be designed to minimise, tolerate, reduce the severity of, or prevent errors. Similar to the ideas of *safety in design* (or *safety by design*) that were presented in Chapter 2, one approach to do this is through "defensive design." This involves considering all the ways that equipment end users might misuse the equipment, and then designing it to make this misuse impossible, or at least minimise its consequences. The key in the ignition of a vehicle is a good example of this, where in most instances it cannot be removed whilst the vehicle's engine is running. Similarly, an air filter that can only be fitted in one orientation is a maintenance-related example. The whole notion is related to the Japanese term *poka-yoke*, which broadly means mistake proofing, making equipment fail-safe, or behaviour shaping to prevent common errors and mistakes.

Looking in more depth here, similar to the hierarchy of control that was listed in Chapter 2, generally making mining equipment error tolerant can be achieved on three broad levels; these roughly correspond to primary, secondary, and tertiary safety:

- *Preventing, removing, and eliminating the errors.* One way to prevent mistakes and errors is by designing "behaviour-shaping" constraints into the equipment. One example is not being able to put mobile equipment into a reverse gear whilst the vehicle is moving forward: an operator's behaviour is constrained in this regard, and this type of error is prevented. The above-mentioned design of an air filter to only be fitted in one particular orientation is another good example here.
- *Limiting the occurrence of errors.* This includes, for example, by a confirmation of an action being required, or by training on normal and emergency procedures. A common, and general, example here is with computer file deletion (where an "Are you sure?"–type message is displayed). In the mining domain, making roadways wider so that some degree of lateral lane deviation is possible (combined with berm walls in case of extreme deviations) is an example. Generally, these measures may have some limitations or side effects: for example, computer users often automatically hit the Confirm Deletion key when prompted, and making roadways wider is often more expensive (or not always possible, especially underground).
- *Mitigating and limiting the effects of errors.* The example in the last bullet point about computer file deletion confirmation is often supplemented by the additional safeguard of a recycling bin or trash can from which the deleted file can still be "saved." This allows recovery in case the file was erroneously deleted (often this file is available in the recycling bin for only a couple of days afterwards before

permanent deletion). The initial error still occurs, but its effects are recoverable within a certain period of time. In the mining domain, as noted in Chapter 3, fuel cut-offs (or deadman's controls) once an incident has occurred can at least minimise its severity. As with the "limiting the occurrence of errors" category, generally many of these measures have some issues: for example, deadman's controls may overload the operator by requiring another action to be performed (i.e., holding the deadman's handle) at moments of high workload (e.g., in emergency situations).

Perhaps the key for future mining equipment is to design and build a multilayered defence that focuses on preventing, reducing, detecting, identifying, recovering/resuming, and limiting the effects of errors to better support human performance (Wood and Kieras, 2002). Similar to the Simpson et al. (2009) quote earlier in this section, they state, "Human error is pervasive. Designing for human error should also be pervasive" (Wood and Kieras, 2002, p. 11).

It is anticipated that further work will be undertaken here by mining equipment manufacturers and designers, and a better dialogue between OEMs and mine sites will help develop additional site-level controls (including training) when and where needed for more comprehensive future equipment error management systems.

12.3.3 Design maturity

Journey models are becoming an increasingly popular way of considering many issues in mining, including organisational issues, safety management, and the design of equipment (Simpson et al., 2009). Focusing specifically on equipment, the maturity of the design process or the maturity of the OEM and customer engagement can often be thought of as a sequential, or step-wise, progression towards a more mature and innovative end point. Looking at this in a more concrete way, Joy (2009) proposed an OEM and customer design maturity chart; a slightly modified version of this is shown in Figure 12.1. This shows five stages that must be reached before attaining a mature and resilient OEM and mining customer design process. Generally it is considered that to progress to higher levels, steps cannot be skipped. Similarly, it should be noted that each step is not equal in the workload or time frame required to achieve progression to the next step.

Whether or not all the exact details in the above figure are accepted, it should be at least broadly accepted that individual mine sites, mining companies as a whole, OEMs, and the OEM–mining customer engagement processes are of varying levels of maturity, particularly regarding safety issues.

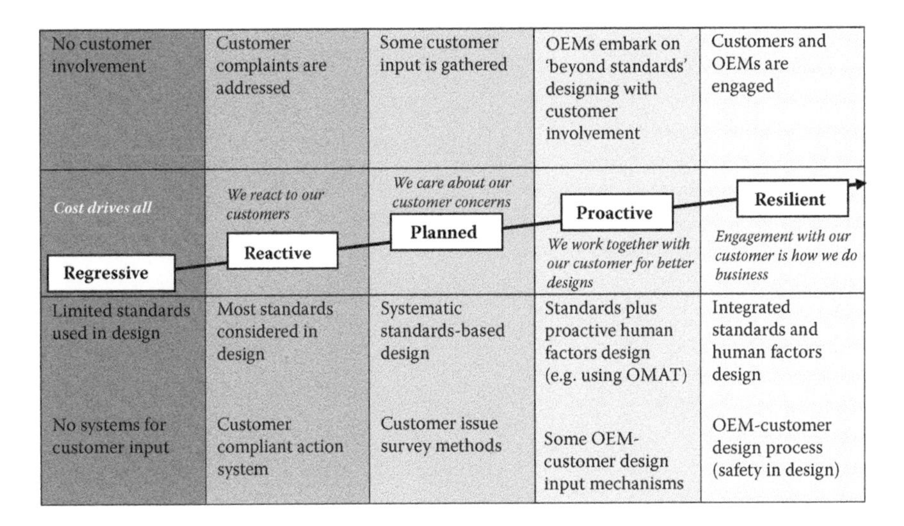

Figure 12.1 OEM and customer design maturity chart. (Adapted from Joy, 2009)

As can be seen in Figure 12.1, human factors considerations are of key importance here. The progression up the maturity chart concept is not a quick journey, but it is anticipated that the use of the material contained in this book will ultimately help mining equipment to be designed, operated, and maintained in a safer and more productive manner. A systematic consideration of the human element when interacting with mining equipment has the potential to be a major step forward in the industry. It is hoped that the material in this book goes some way to helping meet that objective.

References

Abernethy, B. 2001. "Learning from experts: How the study of expertise might help design more effective training." In *Proceedings of the 37th Annual Conference of the Ergonomics Society of Australia*, edited by M. Stevenson and J. Talbot, 3–12. Canberra: Ergonomics Society of Australia.

American Conference of Governmental Industrial Hygienists (ACGIH). 2009. *Threshold Limit Values for Chemical Substances and Physical Agents and Biological Exposure Indices*. Cincinnati, OH: ACGIH.

Bainbridge, L. 1987. "Ironies of automation." In *New Technology and Human Error*, edited by J. Rasmussen, K. Duncan, and J. Leplat, 775–779. Chichester, UK: Wiley. (Reprint of a 1983 Automatica paper 19)

Bell, S. 2009. *Collision Detection Technology Overview*. www.dme.qld.gov.au/zone_files/mines_safety-health/deedi_2009_proximity_workshop_ppt_p1-18.pdf (accessed February 26, 2010).

Bise, C.J. 1997. "Virtual reality: Emerging technology for training of miners." *Mining Engineering* 49 (1): 37–41.

Blignaut, C.J.H. 1979. "The perception of hazard. II. The contribution of signal detection to hazard perception." *Ergonomics* 22:1177–1185.

Boileau, P.E., Boutin, J., Eger, T., and Smets, M. 2006. "Vibration spectral class characterization of load-haul-dump mining vehicles and seat performance evaluation." In *Proceedings of the First American Conference on Human Vibration*, edited by R. Dong, K. Krajnak, O. Wirth, and J. Wu, 14–15. www.cdc.gov/niosh/docs/2006-140/pdfs/2006-140.pdf (accessed February 26, 2010).

Boldt, C.M.R., and Backer, R.R. 1999. "Surface haulage truck research." *American Journal of Industrial Medicine* 36:66–68.

Boocock, M.G., and Weyman, A.K. 1994. "Task analysis applied to computer-aided design for evaluating driver visibility." In *Proceedings of the 12th IEA Triennial Congress*. Toronto, 4:261–263.

Bovenzi, M., and Hulshof, C.T.J. 1998. "An updated review of epidemiologic studies of the relationship between exposure to whole body vibration and low back pain." *Journal of Sound and Vibration* 215:595–611.

Brake, D. 2002. "Design of the world's largest bulk air cooler for the Enterprise Mine in Northern Australia." In *Mine Ventilation: Proceedings of the North American/Ninth US Mine Ventilation Symposium*, edited by E. De Souza, Kingston, Canada, 381–390.

Brake, D. and Bates, G.P. 2000. "Occupational treat illness: An interventional study." In *Proceedings of the International Conference on Physiological and Cognitive Performance in Extreme Environments*, pp 170–172. edited by W.M. Lau. Canberra, Australia.

Brake, D., and Bates, G.P. 2002b. "Limiting metabolic rate (thermal work limit) as an index of thermal stress." *Applied Occupational and Environmental Hygiene* 17 (3):176–186.

Bureau of Labor Statistics. 1998. Occupational injuries and illnesses: Counts, rates, and characteristics, 1995. Washington, DC: US Department of Labor, Bulletin 2493.

Burgess-Limerick, R. 2008. *Procedure for Managing Injury Risks Associated with Manual Tasks*. http://burgess-limerick.com/download/manualtasksprocedure.pdf (accessed February 26, 2010).

Burgess-Limerick, R. 2009. *Principles for the Reduction of Errors in Bolting Control Operation*. ACARP project C16013 final report. Brisbane, Australia: Australian Coal Association Research Program.

Burgess-Limerick, R., Krupenia, V., Zupanc, C., Wallis, G., and Steiner, L. (2010a). Reducing control section errors associated with underground botting equipment. *Applied Ergonomics* 41, 549–555.

Burgess-Limerick, R., Krupenia, V., Wallis, G., Pratim-Bannerjee, A. & Steiner, L. 2010b. Directional control-response relationships for mining equipment. Ergonomics, in press.

Burgess-Limerick, R., Mon-Williams, M., and Coppard, V. 2000. "Visual display height." *Human Factors* 42:140–150.

Burgess-Limerick, R., and Steiner, L. 2006. "Injuries associated with continuous miners, shuttle cars, load-haul-dump, and personnel transport in New South Wales underground coal mines." *Mining Technology* (TIMM A) 115:160–168.

Burgess-Limerick, R., Straker, L., Pollock, C., Dennis, G., Leveritt, S., and Johnson, S. 2007. "Participative ergonomics for manual tasks in coal mining." *International Journal of Industrial Ergonomics* 37:145–155.

Cannon-Bowers, J.A., and Salas, E. 1998. *Making Decisions UNDER Stress: Implications for Individual and Team Training*. Washington, DC: APA Press.

Cantwell, V. 1997. "Physiological factors affecting safety in maritime operations." Paper presented at Safety at Sea International conference, Baltimore, April 30–May 1.

Caretta, T.R., and Dunlap, R.D. 1998. *Transfer of Effectiveness in Flight Simulation: 1986 to 1997*. Mesa, AZ: US Air Force Research Laboratories, National Technical Information Service.

Cecala, A.B., O'Brien, A.D., Pollock, D.E., Zimmer, J.A., Howell, J.E. and McWilliams, L.J. 2007. Reducing respirable dust exposure of workers using an improved clothes cleaning process. *International Journal of Mining Research Eng*, 12(2):73–94.

Cecala, A.B., Organiscak, J., Zimmer, J., Heitbrink, W., Moyer, E., Schmitz, M., Ahrenholtz, E., Coppock, C., Andrews, C. 2005. "Reducing enclosed cab drill operators respirable dust exposure with effective filtration and pressurization techniques. *Journal of occupational and Environmental Hygiene* 2:65–63.

Chan, W.H., and Chan, A.H.S. 2003. "Movement compatibility for rotary control and circular display-computer simulated test and real hardware test." *Applied Ergonomics* 34:61–67.

Chapanis, A. 1996. *Human Factors in Systems Engineering*. New York: Wiley.

Chapanis, A. 1999. *The Chapanis Chronicles: 50 Years of Human Factors Research, Education and Design*. Santa Barbara, CA: Aegean.

Churchill, E., and Snowoten, D. 1996. Mine Training. Interval Project Report. University of Nottingham, UK.

Cliff, D., and Horberry, T. 2008. "Hours of work risk factors for coal mining." *International Journal of Mining and Mineral Engineering* 1 (1): 74–94.

Committee on Technologies for the Mining Industry, National Research Council. 2002. *Evolutionary and Revolutionary Technologies for Mining*. Washington, DC: National Academy Press.

Commonwealth Scientific and Industrial Research Organisation (CSIRO). 2009. *Controlling Mines from a Distance*. www.csiro.au/science/Mine-control.html (accessed February 27, 2010).

Crandall, B., Klein, G., and Hoffman, R. 2006. *Working Minds: A Practitioners Guide to Cognitive Task Analysis*. Cambridge: Massachusetts Institute of Technology.

Crooks, W.W., Drake, K.L., Perry, T.J., Schwalm, N.D., Shaw, B.F., and Stone, B.R. 1980. *Analysis of Work Areas and Tasks to Establish Illumination Needs in Underground Metal and Nonmetal Mines*, vol. 1, Contract J0387230, Bureau of Mines OFR 111(1)-81. Woodland Hills, CA: Perceptronics.

Dainoff, M., Gordon, C., Robinette, K., and Strauss, M. 2004. *Guidelines for Using Anthropometric Data in Product Design*. Santa Monica, CA: HFES.

Daltroy, L.H., Iversen, M.D., Larson, M.G., Lew, R., Wright, E., Ryan, J., Zwerling, C., Fossel, A.H., and Liang, M.H. 1997. "A controlled trial of an educational program to prevent low back injuries." *New England Journal of Medicine* 337:322–328.

Daniels, G.S. 1952. *The Average Man*, TN-WCRD 53-7. Columbus, OH: Wright Patterson Air Force Base. http://www.dtic.mil/cgi–bin/GetTRDOC?AD=AD0102036Location=U2&doc=GetTRDOC.pdf (accessed April 21, 2010).

Dekker, S. 2006. *The Field Guide to Understanding Human Error*. Aldershot, UK: Ashgate.

Delabbio, F.C., Dunn, P.G., Iturregui, L., and Hitchcock, S. 2003. "The application of 3D CAD visualization and virtual reality in the mining and mineral processing industry." Paper presented at the Computer Applications in the Minerals Industries conference, September, Calgary.

Denby, B., Schofield, D., McClarnon, D.J., Williams, M., and Walsha, T. 1998. "Hazard awareness training for mining situations using virtual reality." APCOM 27th International Symposium on Computer Applications in the Minerals Industries, London, 695–705.

Department of Mineral Resources. 2004. *Safety Alert 04-07: Operator Killed Changing Tyre*. www.dpi.nsw.gov.au/__data/assets/pdf_file/0008/66860/Safety-Alert-04-07-operator-killed-changing-tyre1.pdf (accessed February 26, 2010).

Dezelic, V., Apel, D.B., Denney, D.B., Schneider, A.J., Hilgers, M.G., and Grayson, L.R. 2005. "Training for new underground rockbolters using virtual reality." Paper presented at the Sixth International Conference on Computer Applications in the Minerals Industries (CAMI) conferences, September, Banff, Canada.

Dhillon, B.S. 2008. *Mining Equipment Reliability, Maintainability, and Safety*. New York: Springer Verlag.

Donoghue, A.M. 2004. "Heat illness in the US mining industry." *American Journal of Industrial Medicine* 45:351–356.

Donoghue, A.M., Sinclair, M.J., and Bates, G.P. 2000. "Heat exhaustion in a deep underground metalliferous mine." *Occupational Environmental Medicine* 57:165–174.

Driskell, J.E., and Johnston, J.H. 1998. Stress experience training. In J.A. Cannon-Bowers & E. Salas (Eds). Making decisions under stress: Implications for Individual and team training. Washington DC: APA. 191–217.

Eger, T., Salmoni, A., Cann, A., and Jack, R. 2006. "Whole-body vibration exposure experienced by mining equipment operators." *Occupational Ergonomics* 6:121–127.

Eger, T., Stevenson, J., Callaghan, J.P., Grenier, S., and Vib, R.G. 2008. "Predictions of health risks associated with the operation of load-haul-dump mining vehicles: Part 2: Evaluation of operator driving postures and associated postural loading." *International Journal of Industrial Ergonomics* 38: 801–815.

Eger, T.R., Salmoni, A.W., and Whissell, R. 2004. "Factors influencing load-haul-dump operator line of sight in underground mining." *Applied Ergonomics* 35:93–103.

Endsley, M.R., Bolte, B., and Jones, D.G. 2003. *Designing for Situation Awareness.* London: Taylor & Francis.

Ericsson, K.A., Krampe, R.T., and Tesch-Romer, C. 1993. "The role of deliberate practice in the acquisition of expert performance." *Psychological Review* 100:363–406.

European Union. 2002. Directive 2002/44/EC of the European Parliament and of the Council. Brussels: European Union. http://eur-lex.europa.eu/LexUriServ/LexUriServ.do?uri=OJ:L:2002:177:0013:0019:EN:PDF (accessed February 28, 2010).

Eye Diseases Prevalence Research Group (EDPRG). 2004. "The prevalence of refractive errors among adults in the United States, Western Europe, and Australia." *Archives of Ophthalmology* 122:495–505.

Feare, T. 1999. "Forklift operator training: What OSHA's new rules require you to do." *Modern Materials Handling* 64, 40–43.

Filigenzi, M.T., Orr, T.J., and Ruff, T.M. 2000. "Virtual reality for mine safety training." *Applied Occupational and Environmental Hygiene* 15 (6):465–469.

Fitts, P.M. (ed.). 1951. *Human Engineering for an Effective Air-Navigation and Traffic-Control System.* Columbus: Ohio State University Research Foundation.

Gallagher, S., Mayton, A.G., Unger, R.L., Hamrick, C.A., and Sonier, P. 1996. "Computer design and evaluation tool for illuminating underground coal mining equipment." *Journal of the Illuminating Engineering Society* 25 (1):3–12.

Gilbert, V.A. 1990. *Research Support for the Development of SAE Guidelines for Underground Operator Compartments*, USBM OFR 8–91. Washington, DC: US Bureau of Mines.

Gilhooley, K.J., and Green, A.J.K. 1988. "The use of memory by experts and novices." In *Cognition and Action in Skilled Behavior*, edited by A.M. Colley and J.R. Beech, 379–395. Amsterdam: North-Holland.

Glaser, R., and Chi, M.T.H. 1988. "Overview." In *The Nature of Expertise*, edited by M.T.H. Chi, R. Glaser, and M.J. Farr, xv–xxvii. Hillsdale, NJ: Erlbaum.

Godwin, A., and Eger, T. 2009. "Using virtual computer analysis to evaluate the potential use of a camera intervention on industrial machines with line-of-sight impairments." *International Journal of Industrial Ergonomics* 39 (1): 146–151.

Godwin, A., Eger, T., Salmoni, A., and Dunn, P. 2008. "Virtual design modifications yield line-of-sight and safety related improvements for load-haul-dump operators." *International Journal of Industrial Ergonomics* 38:202–210.

Gordon, S.E. 1994. *Systematic Training Program Design: Maximising Effectiveness and Minimizing Liability*. Englewood Cliffs, NJ: Prentice Hall.

Grech, M., Horberry, T., and Koester, T. 2008. *Human Factors in the Maritime Domain*. Boca Raton, FL: CRC Press.

Griffin, M.J. 1990. *Handbook of Human Vibration*. London: Academic Press.

Griffin, M.J. 1998. "Evaluating the effectiveness of gloves in reducing the hazards of hand-transmitted vibration." *Occupational and Environmental Medicine* 55:340–348.

Griffin, M.J. 2006. "Vibration and motion." In *Handbook of Human Factors and Ergonomics*, 3rd ed., edited by G. Salvendy. Hoboken, NJ: Wiley.

Gurusamy, K., Aggarwal, R., Palanivelu, L., and Davidson, B.R. 2008. "Systematic review of randomized controlled trials on the effectiveness of virtual reality training for laprascopic surgery." *British Journal of Surgery* 95 (9): 1088–1097.

Haddon, W.A. 1972. "A logical framework for categorizing highway safety phenomena and activity." *Journal of Trauma* 12:193–207.

Haddon, W.A. 1973. "Energy damage and the 10 countermeasure strategies." *Journal of Trauma* 13:321–331.

Hall, R.H., Nutakor, D., Apel, D., Grayson, L., Hilgers, M.G., and Warmbolt, J. 2008. "Evaluation of a virtual reality simulator developed for training miners to install rock bolts using a Jackleg drill." In *Proceedings of the Annual Conference of the Society for Mining Engineers*. Preprint No. 08–054.

Hartley, L., Horberry, T., Mabbott, N., and Krueger, G.P. 2000. *Review of Fatigue Detection and Prediction Technologies*, Technical Report. Melbourne, Australia: National Road Transport Commission.

Haslam, C., Clemes, S., McDermott, H., Shaw, K., Williams, C., and Haslam, R. 2007. *Manual Handling Training: Investigation of Current Practices and Development of Guidelines*. www.hse.gov.uk/research/rrhtm/rr583.htm (accessed February 26, 2010).

Hays, R.T., Jacobs, J.W., Prince, C., and Salas, E. 1992. "Flight simulator effectiveness: A meta-analysis." *Military Psychology* 4 (2): 63–74.

Hedlund, U. 1989. "Raynaud's phenomenon of fingers and toes of miners exposed to local and whole-body vibration and cold." *International Archives of Occupational and Environmental Health* 61:457–461.

Helander, M.G., Conway, E.J., Elliott, W., and Curtin, R. 1980. *Standardization of Controls for Roof Bolter Machines: Phase 1: Human Factors Engineering Analysis*, USBM OFR 170-82 PB83-119149. Washington, DC: US Bureau of Mines.

Helander, M.G., Krohn, G.S., and Curtin, R. 1983. "Safety of roof-bolting operations in underground coal mines." *Journal of Occupational Accidents* 5:161–175.

Hendrick, H.W. 2003. "Determining the cost-benefits of ergonomics projects and factors that lead to their success." *Applied Ergonomics* 34 (5): 419–427.

Hoffmann, E.R. 1997. "Strength of component principles determining direction of turn stereotypes: Linear displays with rotary controls." *Ergonomics* 40:199–222.

Horberry, T., and Edquist, J. 2008. "Distractions outside the vehicle." In *Driver Distraction*, edited by M. Regan, J.D. Lee, and K. L. Young. 215–217. Boca Raton, FL: CRC Press.

Horberry, T., Gunatilaka, A., and Regan, M. 2006. "Intelligent systems for industrial mobile equipment." *Journal of Occupational Health and Safety: Australia and New Zealand* 22 (4): 323–334.

Horberry, T., Hutchins, R., and Tong, R. 2008. *Motorcycle Rider Fatigue: A Review*, Department for Transport, Road Safety Research Report no. 78. www.dft.gov.uk/pgr/roadsafety/research/rsrr/theme2/riderfatigue.pdf (accessed February 26, 2010).

Horberry, T., Larsson, T., Johnston, I., and Lambert, J. 2004. "Forklift safety, traffic engineering and intelligent transport systems: A case study." *Applied Ergonomics* 35 (6): 575–581.

Horberry, T., Regan, M., and Toukhsati, S. 2007. "Airport ramp safety and intelligent transport systems." *IET Intelligent Transport Systems* 1 (4): 234–240.

Horberry, T., Sarno, S., Cooke, T., and Joy, J. 2009. *Development of the Operability and Maintainability Analysis Technique for Use with Large Surface Haul Trucks*, Australian Coal Association Research Program report. www.acarp.com.au/abstracts.aspx?repId=C17033 (accessed February 26, 2010).

Hoy, J., Mubarak, N., Nelson, S., Sweerts de Landas, M., Magnusson, M., Okunribido, O., and Pope, M. 2005. "Whole body vibration and posture as risk factors for low back pain among forklift truck drivers." *Journal of Sound and Vibration* 284:933–946.

Human Factors and Ergonomics Society of Australia. 2010. *About the HFESA*. www.ergonomics.org.au/about.aspx (accessed February 26, 2010).

Human Solutions. n.d. [Home page]. www.human-solutions.com (accessed February 27, 2010).

Humphries, M. 1958. "Performance as a function of control-display relations, positions of the operator, and locations of the control." *Journal of Applied Psychology* 42:311–316.

Illuminating Engineering Society of North America (IES). 1993. *Lighting Handbook: Reference and Application*, 8th ed. New York: IES.

Industry & Investment NSW. 2010. Guideline for Bolting and Drilling Plant in Mines: Part1: Bolting Plant for Strata Support in Underground Coal Mines, MDG 35.1. Sydney: I&I NSW.

International Organization for Standardization (ISO). 1997. *Mechanical Vibration and Shock: Evaluation of Human Exposure to Whole-Body Vibration, Part 1: General Requirements*, ISO 2631-1. Geneva: ISO.

International Organization for Standardization (ISO). 2001. *Mechanical Vibration: Measurement and Evaluation of Human Exposure to Hand-Transmitted Vibration, Part 1: General Requirements*, ISO 5349-1. Geneva: ISO.

International Organization for Standardization (ISO). 2004. *Mechanical Vibration and Shock: Evaluation of Human Exposure to Whole-Body Vibration: Part 5: Method for Evaluation of Vibration Containing Multiple Shocks*, ISO 2631-5. Geneva: ISO.

International Organization for Standardization (ISO). 2008. *Earth-Moving Machinery: Determination of Sound Power Level: Dynamic Test Conditions*, ISO 6395. Geneva: ISO.

Jacobs, J.W., Prince, C., Hays, R.T., and Salas, E. 1990. *A Meta-analysis of the Flight Simulator Training Research*, NAVTRASYSCEN TR-89-006. Orlando, FL: Human Factors Division, Naval Training Systems Center.

Joy, J. 2009. "Introduction to the EMESRT OEM design maturity chart." Personal communication.

Kielblock, A.J., and Schutte, P.C. 1993. "Human heat stress: Basic principles, consequences and its management." In *MINESAFE International 1993*, Perth, Australia, March 21–26, 279–294.

Kihl, M., and Wolf, P.J. 2007. "Using driving simulators to train snowplow operators: The Arizona experience." In *Proceedings of the 2007 Mid-Continent Transportation Research Symposium*, Ames, Iowa, August 16–17. www.intrans.iastate.edu/pubs/midcon2007/KihlSnowplow.pdf (accessed February 26, 2010).

Kittusamy, N., and Buchholz, B. 2004. "Whole-body vibration and postural stress among operators of construction equipment: A literature review." *Journal of Safety Research* 35:255–61.

Kizil, M. 2003. *Virtual Reality Applications in the Australian Minerals Industry: Application of Computers and Operations Research in the Minerals Industries*. Marshalltown: South African Institute of Mining and Metallurgy.

Klishis, M.J., Althouse, R.C., Stobbe, T.J., Plummer, R.W., Grayson, R.L., Layne, L.A., and Lies, G.M. 1993. *Coal Mine Injury Analysis: A Model for Reduction through Training: Volume VIII: Accident Risks during the Roof Bolting Cycle: Analysis of Problems and Potential Solutions*, USBM Cooperative agreements C0167023 and C0178052. Morgantown: West Virginia University.

Kowalski-Trakofler, K.M., Vaught, C., and Scharf, T. 2003. "Judgment and decision making under stress: An overview for emergency managers." *International Journal of Emergency Management* 1 (3):278–289.

Kroemer, K.H.E., and Grandjean, E. 1997. *Fitting the Task to the Human*, 5th ed. London: Taylor & Francis.

Kryter, K.D. 1994. *The Handbook of Hearing and the Effects of Noise*. San Diego: Academic Press.

Kumar, S. 2004. "Vibration in operating heavy haul trucks in overburden mining." *Applied Ergonomics* 35:509–520.

Larsson, T.J., and Rechnitzer, G. 1994. "Forklift trucks: Analysis of severe and fatal occupational injuries, critical incidents and priorities for prevention." *Safety Science* 17:275–289.

Lemyre, L., and Tessier, R. 2003. *Measuring Psychological Stress: Concept, Model, and Measurement Instrument in Primary Care Research*. www.cfpc.ca/cfp/2003/Sep/vol49-sep-resources-6.asp (accessed February 26, 2010).

Lemyre, L., Tessier, R., and Fillion, L. 1990. *La Mesure du Stress Psychologique: Manuel d'Utilisation*. Québec: Behaviora.

Leveritt, S. 1998. *Heat Stress in Mining*. www.uq.edu.au/eaol/apr98/leveritt.pdf (accessed February 26, 2010).

Lindsey, J.T. 2005. "The perceptions of emergency vehicle drivers using simulation in driver training." In *Proceedings of the Third International Driving Symposium on Human Factors in Driver Assessment, Training and Vehicle Design*, June 27–30, Rockport, ME. http://ppc.uiowa.edu/driving-assessment/2005/final/papers/10_Lindseyformat.pdf (accessed February 26, 2010).

Loveless, N.E. 1962. "Direction-of-motion stereotypes: A review." *Ergonomics* 5:357–383.

Lucas, J., and Thabet, W. 2008. "Implementation and evaluation of a VR task-based training tool for conveyor belt safety training." *Journal of Information Technology in Construction* 13:637–659.

Lucas, J., Thabet, W., and Worlikar, P. 2008. "A VR-based training program for conveyor belt safety." *Journal of Information Technology in Construction* 13:381–407.

Mark, C. 2002. "The introduction of roof-bolting to U.S. underground coal mines (1948–1960): A cautionary tale." In *Proceedings of 21st International Conference on Ground Control in Mining*. Morgantown: West Virginia University, 150–160.

Martimo, K.-P., Verbeek, J.H., Karppinen, J., Furlan, A.D., Kuijer, P.P.F.M., Viikari-Juntura, E., Takala, E.-P., and Jauhiainen, M. 2007. "Manual material handling advice and assistive devices for preventing and treating back pain in workers." *Cochrane Database of Systematic Reviews* 2006 (2). www.cochrane.org/reviews/en/ab005958.html (accessed February 26, 2010).

Martin, R., and Graveling, R. 1983. "Background illumination and its effects on peripheral vision awareness for miners using caplamps." *Applied Ergonomics* 14:139–141.

Marx, K.W. 1987. "Ergonomic factors in LHD design." *Mining*, December, 549–556.

Mason, S., and Rushworth, A.M. 1991. *Improving Mining Machinery Maintainability*. London: Sterling Publications for the British Coal Corporation.

McMahan, R.P., Bowman, D.A., Schafrik, S., and Karmis, M. 2008. "Virtual environment training for preshift inspections of haul trucks to improve mining safety." In *Proceedings of the First International Future Mining Conference*, Sydney, November 19–21, 167–173.

McPhee, B. 2005. *Practical Ergonomics: Application of Ergonomics Principles in the Workplace*. Corrimal, Australia: Coal Services Health and Safety Trust, 48–49.

McPhee, B., Foster, G., and Long, A. 2009. *Bad Vibrations*, 2nd ed. Sydney: Coal Services Health and Safety Trust.

Miller, V.S., and Bates, G.P. 2007. "The thermal work limit is a simple reliable heat index for the protection of workers in thermally stressful environments." *Annals of Occupational Hygiene* 51 (6):553–561.

Miller, W.K., and McLellan, R.R. 1973. *Analysis of Disabling Injuries Related to Roof Bolting in Underground Bituminous Coal Mines: 1973*. US Department of the Interior Informational Report 1107. Washington, DC: US Department of the Interior.

Minerals Industry Safety and Health Centre (MISHC). 2009. *Earth Moving Equipment Safety Round Table*. www.mishc.uq.edu.au/index.html?page=58384 (accessed February 26, 2010).

Mine Safety and Health Administration (MSHA). 1994. *Coal Mine Safety and Health Roof-Bolting-Machine Committee*. Arlington, VA: MSHA.

Mine Safety and Health Administration (MSHA). 1997. *Safety Standards for the Use of Roof-Bolting Machines in Underground Mines*. Arlington, VA: MSHA. www.msha.gov/REGS/FEDREG/PROPOSED/1997PROP/97-32203.HTM (accessed March 6, 2010).

Mine Safety and Health Administration (MSHA). 2001. *Directorate of Technical Support*. Arlington, VA: MSHA. www.msha.gov/Accident_Prevention/appmain.htm (accessed February 26, 2010).

Mines and Aggregates Safety and Health Association (MASHA). 2001. *Ontario Mining Industry Accidents: All Firms 1985–2000 Underground Accidents while Operating LHD Vehicles*. North Bay, ON: MASHA.

MIRMgate. 2010. *Minerals Industries Safety Resource*. www.mirmgate.com (accessed February 26, 2010).

Misagi, F.L. 1976. *Heat Stress in Mining*. Mining Enforcement and Safety Administration (MESA) Safety Manual no. 6. Washington, DC: MESA.

Misagi, F.L., Inderberg, J.D., Blumenstein, P.D., and Naiman, T. 1976. *Heat Stress in Hot US Mines and Criteria for Standards for Mining in Hot Environments*. Mining Enforcement and Safety Admininistration Informational Report no. 1048. Washington, DC: MESA.

Mitchell, M.J.H., and Vince, M.A. 1951. "The direction of movement of machine controls." *Quarterly Journal of Experimental Psychology* 3:24–35.

Mumaw, R.J., and Roth, E.M. 1995. "Training complex tasks in a functional context." In *Proceedings of the Human Factors and Ergonomics Society 39th Annual Meeting* 1253–1257. Santa Monica CA: HFES.

NASA. 1995. *Man-Systems Integration Standards*, NASA-STD-3000. http://msis.jsc.nasa.gov (accessed February 26, 2010).

National Institute for Occupational Safety and Health (NIOSH). 1986. *Criteria for a Recommended Standard: Occupational Exposure to Hot Environments*. US Department of Health and Human Services Publication no. 86–113. Washington, DC: US Department of Health and Human Services.

National Institute for Occupational Safety and Health (NIOSH). 1997. *Musculoskeletal Disorders and Workplace Factors: A Critical Review of Epidemiologic Evidence for Work-Related Musculoskeletal Disorders of the Neck, Upper Extremity, and Low Back*. US Department of Health and Human Services Publication no. 97-141. Washington, DC: US Department of Health and Human Services.

National Institute for Occupational Safety and Health (NIOSH). 2009. *Mining: Occupational Safety and Health Risks*. www.cdc.gov/niosh/programs/mining/risks.html (accessed February 26, 2010).

Newell, G.S., and Mansfield, N.J. (2008). "Evaluation of reaction time performance and subjective workload during whole-body vibration exposure while seated in upright and twisted postures with and without armrests." *International Journal of Industrial Ergonomics* 38:499–508.

New South Wales Department of Primary Industries (NSW DPI). 1995. *Guideline for Free-Steered Vehicles*, MDG 1. Sydney: New South Wales Department of Primary Industries.

New South Wales Department of Primary Industries (NSW DPI). 2002. *Guideline for Mobile and Transportable Equipment for Use in Mines*, MDG 15. Sydney: New South Wales Department of Primary Industries.

NexGen Ergonomics. 2009. [Home page]. www.nexgenergo.com (accessed February 27, 2010).

Nutakor, D. 2008. *Design and Evaluation of a Virtual Reality Training System for New Underground Rockbolters*. PhD thesis, Missouri University of Science and Technology. http://scholarsmine.mst.edu/thesis/pdf/Nutakor_09007dcc80672480.pdf (accessed February 26, 2010).

Organiscak, J.A. and Cecala, A.B. 2008. Key Design factors of Enclosed cab Dust filtration systems. NIOSH Report of Investigation 9677, November, 43 pp.

Orr, T.J., Fligenzi, M.T., and Ruff, T.M. 2008. "Desktop virtual reality miner training simulator." Spokane, WA: NIOSH Research Laboratory.

Parkes, A. M., and Reed, N. 2006. "Transfer of fuel-efficient driving technique from the simulator to the road: Steps towards a cost-benefit model for synthetic training." In *Developments in Human Factors in Transportation, Design, and Evaluation*, edited by D. de Waard, K.A. Brookhuis, and A. Toffetti, 163–176. Maastricht: Shaker.

Patterson, J. 2008. The Development of an Accident/Incident Investigation system for the Mining Industry based on the Human Factors Analysis and Classification System (HFACS) framework. Paper presented at the 2008 Queensland Mining Industry Health & Safety Conference, Townsville, Australia, August 2008. Accessed April 1 2010 from http:www.qrc.org.au/conference/-dbase-upl.papers2008_patterson.pdf

Peebles, L., and Norris, B. 2003. Filling 'gaps' in strength data for design. *Applied Ergonomics* 34, 73–88.

PeopleSize. 2008. *PeopleSize 2008*. www.openerg.com/psz/ (accessed February 27, 2010).

Peters, E., Lipkus, I., and Diefenbach, M.A. 2006. "The functions of affect in health communications and in the construction of health preferences." *Journal of Communication* 56 (1):140–150.

Pfeiffer, M.G., Horey, J.D., and Butrimas, S.K. 1991. "Transfer of simulated instrument training to instrument and contact flight." *International Journal of Aviation Psychology* 1:219–229.

Pheasant, S., and Haslegrave, C.M. 2006. *Bodyspace: Anthropometry, Ergonomics and the Design of the Work*, 3rd ed. Boca Raton, FL: Taylor & Francis.

Pope, M., Wilder, D., and Magnusson, M. 1998. "Possible mechanisms of low back pain due to whole-body vibration." *Journal of Sound and Vibration* 215:687–697.

Poplin, G.S., Miller, H.B., Ranger-Moore, J., Bofinger, C.M., Kurzius-Spencer, M., Harris, R.B., and Burgess, J.F. 2008. "International comparison of injury rates in coal mining: A comparison of risk and compliance-based regulatory approaches." *Safety Science* 46:1196–1204.

Queensland Government Mines Inspectorate, Safety and Health Division. 2001. *Dozer Runs over Light Vehicle*. www.dme.qld.gov.au/zone_files/inspectorate_pdf/incident_report038.pdf (accessed February 26, 2010).

Queensland Mines and Quarries. 2005. *Queensland Mines and Quarries Safety Performance and Health Report 1 July 2004 to 30 June 2005*. www.dme.qld.gov.au/zone_files/publications/annual_report0405.pdf (accessed February 26, 2010).

Queensland Mines and Quarries. 2008. *Queensland Mines and Quarries Safety Performance and Health Report 1 July 2007 to 30 June 2008*. www.energy.qld.gov.au/zone_files/publications/109008_minesco_safety_rpt.pdf (accessed February 26, 2010).

Ramsey, J.D., Burford, C.L., Beshir, M.Y., and Jensen, R.L. 1982. "Effects of workplace thermal conditions on safe work behaviour." *Journal of Safety Research* 3:105–114.

Rasche, T. 2009. *Collision Detection Technology Review*. www.dme.qld.gov.au/zone_files/mines_safety-health/deedi_2009_proximity_workshop_butcherpaper_notes.pdf (accessed February 26, 2010).

Regan, M.A., Lee, J.D., and Young, K.L. 2008. *Driver Distraction: Theory, Effects, and Mitigation.* Boca Raton, FL: CRC Press.

Reyes, M.A., Gallagher, S., and Sammarco, J.J. 2009. Evaluation of visual performance when using incandescent, flourescent, and LED machine lights in mesopic conditions. IAS Conference Record of the 2009 IEEE Industry Applications Conference: Houston, Texas.

Rider, J.P and Colinet, J.F. 2007. Current Dust Control Practices on U.S. Longwalls. Longwall USA: Pittsburgh, PA.

Riordan, C.A., Johnson, G.D., and Thomas, J.S. 1991. "Personality and stress at sea." *Journal of Social Behaviour and Personality* 6 (7): 391–409.

Roberts, J., and Corke, P. 2000. *Obstacle Detection for a Mining Vehicle Using a 2D Laser.* In *Proceedings of the Australian Conference on Robotics and Automation,* Melbourne, Australia 185–190.

Robinette, K.M., and Hudson, J. 2006. "Anthropometry." In *Handbook of Human Factors and Ergonomics,* edited by G. Salvendy, 322–339. New York: John Wiley.

Roscoe, S.N. 1980. *Aviation Psychology.* Ames: Iowa State University Press.

Rouse, W.B., and Boff, K.R. 1997. "Assessing cost/benefit of human factors." In *Handbook of Human Factors and Ergonomics,* 2nd ed. edited by G. Salvendy. New York: John Wiley.

Ruff, T.M. 2001a. *Miner Training Simulator: User's Guide and Scripting Language/ Documentation,* DHHS Publication no. 2001-136. Pittsburg, PA: NIOSH, US Department of Health and Human Services.

Ruff, T.M. 2001b. "Monitoring blind spots: A major concern for haul trucks." *Engineering Mining Journal* 202:17–26.

Rushworth, A.M. 1996. "Reducing accident potential by improving the ergonomics and safety of locomotive and FSV driver cabs by retrofit." *Mining Technology* 78 (898):153–159.

Safe-Away N.d. Safe-Away Ladder & Stairway access systems. www.safe-away.com.au (accessed March 31, 2010).

Safe Work Australia. 2009. *What Is Safe Design?* www.safeworkaustralia.gov.au/swa/HealthSafety/SafeDesign/Understanding/Whatissafedesign.htm (accessed February 26, 2010).

Salas, E., and Cannon-Bowers, J. 2001. "The science of training: A decade of progress." *Annual Reviews of Psychology* 52:471–499.

Sammarco, J.J., Fisher, T.J., Welsh, J.H., and Pazuchanics, M.J. 2001. *Programmable Electronic Mining Systems: Best Practice Recommendations (in Nine Parts),* NIOSH Publications, IC 9456. Pittsburgh, PA: National Institute for Occupational Safety and Health.

Sammarco, J., Reyes, M.A., Freyssinier, J.P., Bullough, J.D., and Zhang, X. 2009a. "Technological aspects of solid-state and incandescent sources for miner cap lamps." *IEEE Transactions Industrial Applications* 45(5):1583–1588.

Sammarco, J., Reyes, M., and Gallagher, S. 2009b. Annual Meeting and Exhibit, February 22–25, Denver, Co. Preprint No. 09–090. Littleton, Co. *Society for Mining, Metallurgy and Exploration,* 2009: 1–5.

SAMMIE CAD. 2004. *Welcome to SAMMIE CAD.* www.lboro.ac.uk/departments/cd/research/groups/erg/sammie/home.htm (accessed February 27, 2010).

Sanders, M.S., and Peay, J.M. 1988. *Human Factors in Mining,* no. IC 9182. Pittsburgh, PA: U.S. Department of the Interior, Bureau of Mines.

Scharf, T., Vaught, C., Kidd, P., Steiner, L., Kowalski, K., Wiehagen, W., Rethi, L., and Cole, H. 2001. "Toward a typology of dynamic and hazardous work environments." *Human and Ecological Risk Assessment* 7(7):1827–1841.

Schmidt, R.A., and Bjork, R.A. 1992. "New conceptualizations of practice: Common principles in three paradigms suggest new concepts for training." *Psychological Science* 3:207–217.

Schofield, D. 2008. "Do you learn more when your life is in danger? How to successfully utilize virtual simulators in the minerals industries." In *Proceedings of the First International Future Mining Conference*, November 19–21, Sydney, 183–188.

Schofield, D., Hollands, R., and Denby, B. 2001. "Mine safety in the twenty-first century: The application of computer graphics and virtual reality." In *Mine Health and Safety Management*, edited by M. Karmis, 153–174. Littleton, CO: Society for Mining, Metallurgy, and Exploration.

Schweigert, M. 2002. "The relationship between hand-arm vibration and lower extremity clinical manifestations: A review of the literature." *International Archive Occupational Environment and Health* 75:179–185.

Sheridan, T. 2002. *Humans and Automation*. New York: John Wiley.

Siemens. 2010. *JACK and Process Simulate Human*. www.plm.automation.siemens.com/en_us/products/tecnomatix/assembly_planning/jack/index.shtml (accessed February 27, 2010).

Silverstein, B., and Clark, R. 2004. "Interventions to reduce work-related musculoskeletal disorders." *Journal of Electromyography and Kinesiology* 14:135–152.

Simpson, G.C., and Chan, W.L. 1988. "The derivation of population stereotypes for mining machines and some reservations on the general applicability of published stereotypes." *Ergonomics* 31:327–335.

Simpson, G., Horberry, T., and Joy, J. 2009. *Understanding Human Error in Mine Safety*. Surrey, UK: Ashgate.

Skogsberg, L. 2006. "Vibration control on hand-held industrial power tools." In *Proceedings of the First American Conference on Human Vibration*, edited by R. Dong, K. Krajnak, O. Wirth, and J. Wu, 108–109. Washington, DC: NIOSH. www.cdc.gov/niosh/docs/2006-140/pdfs/2006-140.pdf (accessed February 26, 2010).

Society of Automotive Engineers International (SAE International). 2010. *Civilian American and European Surface Anthropometry Research Project (CAESAR)*. http://store.sae.org/caesar/ (accessed February 26, 2010).

Squelch, A.P. 1997. *Virtual Reality Simulators for Rock Engineering Related Training*, SIMRAC project GAP 420 final report. Pretoria, South Africa: CSIR Mining Technology.

Standards Australia. 1988. *1988 Vibration and Shock: Hand-Transmitted Vibration: Guidelines for Measurement and Assessment of Human Exposure*, AS 2763-1988. Sydney: Standards Australia.

Standards Australia. 1991. *Earth-Moving Machinery: Design Guide for Access Systems*, AS 3868-1991. Sydney: Standards Australia.

Standards Australia. 1992. *Fixed Platforms, Walkways, Stairways and Ladders: Design, Construction and Installation*, AS 1657-1992. Sydney: Standards Australia.

Standards Australia. 2001. *2001 Evaluation of Human Exposure to Whole Body Vibration, Part 1: General Requirements*, AS 2670.1-2001. Sydney: Standards Australia.

Standards Australia. 2005. *Occupational Noise Management*, AS 1269-2005. Sydney: Standards Australia.

Standards Australia. 2006. *Part 1903: Displays, Controls, Actuators and Signals: Ergonomic Requirements for the Design of Displays and Control Actuators: General Principles for Human Interactions with Displays and Control Actuators: Control Actuators*, AS4024 (Safety of machinery). Sydney: Standards Australia.

Starkes, J.L., and Lindley, S. 1994. "Can we hasten expertise by video simulations?" *Quest* 46:211–222.

Steiner, L., Cornelius, K., and Turin, F. 1999. "Predicting systems interactions in the design process." *American Journal of Industrial Medicine* 36: 58–60.

Stothard, P.M., Galbin, J.M., and Fowler, J.C.W. 2004. "Development, demonstration, and implementation of a virtual reality simulation capability for coal mining operations." In *Proceedings ICCR Conference*, Beijing, China.

Stutts, J.C., Reinfurt, D.W., Staplin, L., and Rodgman, E. 2001. *The Role of Driver Distraction in Traffic Crashes.* Washington, DC: AAA Foundation for Traffic Safety.

Swadling, P., and Dudley, J. 2001. "VRLoader: A virtual reality training system for a mining application." Paper presented at SimTecT 2001, Sydney.

Tak, S., Davis, R., and Calvert, G. 2009. "Exposure to hazardous workplace noise and use of hearing protection devices among US workers–NHANES, 1999–2004". *American Journal of Industrial Medicine*, 52(5):358–371.

Thwaites, P. 2008. "Process control in metallurgical plants: Towards operational performance excellence." Plenary talk at the Automining 2008: International Congress in Automation in Mine Industry, Santiago, Chile.

Tichon, J., and Wallis, G. 2009. "Virtual reality stress training and simulator complexity: Why sometimes more is less." *Behaviour and Information Technology.* DOI: 10.1080/014492903420184.

Torma-Krajewski, J., Steiner, L.J., and Burgess-Limerick, R. 2009. *Ergonomics Processes: Implementation Guide and Tools for the Mining Industry*, DHHS (NIOSH) Publication no. 2009-107, Information Circular 9509, February. Pittsburgh, PA: US Department of Health and Human Services, Public Health Service, Centers for Disease Control and Prevention, National Institute for Occupational Safety and Health.

Trotter, D.A. 1982. *The Lighting of Underground Mines*, Trans Tech Publications Series on Mining Engineering no. 2. Zurich: Trans Tech Publications.

Turpin, D., and Welles, R. 2006. "Simulator-based training effectiveness through objective driver performance measurement." Paper presented at the DSC Asia/Pacific, Tsukuba, Japan, May–June. www.teenresearchcenter.org/research%20articles/Simulator%20Based%20Training%20Effectiveness.pdf (accessed February 26, 2010).

UK Department of Trade and Industry. 2000. *Strength Data for Design Safety: Phase 1.* URN 00/1070. DTI: LONDON.

UK Department of Trade and Industry. 2002. Strength Data for Design Safety: Phase 2. URN 02/744. DTI: LONDON.

US Department of Defense. 1991. *Military Handbook: Anthropometry of US Military Personnel*, DOD-HDBK-743A. Washington, DC: US Department of Defense. www.everyspec.com/DoD/DoD-HDBK/ (accessed February 26, 2010).

US Department of Defense. 1999. *Human Engineering*, MIL-STD-1472F. http://safetycenter.navy.mil/instructions/osh/MILSTD1472F.pdf (accessed February 26, 2010).

US Department of Energy. 2009. *Energy Efficiency of White LEDs*. http://apps1.eere.energy.gov/buildings/publications/pdfs/ssl/energy_efficiency_white_leds.pdf (accessed February 26, 2010).

US Federal Aviation Authority (FAA). 2009. *Human Factors Design Guide*. http://hf.tc.faa.gov/hfds/ (accessed February 26, 2010).

Van der Walt, W.A. 1981. *A Survey of the Incidence of Heat Stroke in the Gold Mining Industry over the Period 1969 to 1980*, Report no. 15/81. Johannesburg, South Africa: Chamber of Mines.

Vicente, K. 1999. *Cognitive Work Analysis: Toward Safe, Productive, and Healthy Computer-based Work*. Mahwah, NJ: Lawrence Erlbaum.

Village, J., Morrison, J., and Leong, D. 1989. "Whole-body vibration in underground load-haul-dump vehicles." *Ergonomics* 32 (10): 1167–1183.

Vince, M.A. 1945. *Direction of Movement of Machine Controls*. MRC APU Report number 233. Cambridge, UK.

Vince, M.A., and Mitchell, M.J.H. 1946. "Direction of movement of machine controls II." *Quarterly Journal of Experimental Psychology* 3 (1): 24–35.

Wallis, G.M., Chatziastros, A., and Bülthoff, H.H. 2002. "An unexpected role for visual feedback in vehicle steering control." *Current Biology* 12:295–299.

Weitz, J. 1947. "The coding of airplane control knobs." In *Psychological Research on Equipment Design*, Army Air Forces Aviation Psychology Program Research Report no. 19, edited by P.M. Fitts, 187–198. Washington, DC: Government Printing Office.

Wickens, C.D., Gordon, S.E., and Liu, Y. 1998. *An Introduction to Human Factors Engineering*. New York: Addison Wesley Longman.

Wikström, B. 1993. "Effects from twisted postures and whole-body vibration during driving." *International Journal of Industrial Ergonomics* 12:61–75.

Wilkes, J.T. 2001. "Caterpillar simulation training." In *Proceedings of the Thirty-Second Annual Institute on Mining Health, Safety and Research*, edited by F.M. Jenkins, J. Langton, M.K. McCarter, and B. Rowe, 65–67.

Williams, A.M., and Grant, A. 1999. "Training perceptual skill in sport." *International Journal of Sport Psychology* 30:194–220.

Wood, S.D., and Kieras, D.E. 2002. "Modeling human error for experimentation, training, and error-tolerant design." In *Proceedings of the Interservice/Industry Training, Simulation, and Education Conference*, 1075–1085. December. Orlando, FL.

Worringham, C.J. 2003. "Seriously incompatible: Why some control designs remain lethal." In *Proceedings of the 39th Annual Conference of the Ergonomics Society of Australia*, edited by R. Burgess-Limerick. Ergonomics Society of Australia, Canberra. 12–18.

Worringham, C.J., and Beringer, D.B. 1989. "Operator orientation and compatibility in visuo-motor task performance." *Ergonomics* 32:387–399.

Worringham, C.J., and Beringer, D.B. 1998. "Directional stimulus-response compatibility: A test of three alternative principles." *Ergonomics* 41:864–880.

Wright, N., Stone, B., Horberry, T.J., and Reed, N. 2007. *A Review of In-Vehicle Sleepiness Detection Devices*, TRL Report PPR157. Crowthorne, UK: TRL.

Ye, N., and Salvendy, G. 1996. "Expert-novice knowledge of computer programming at different levels of abstraction." *Ergonomics* 39:461–481.

Zakay, D., and Wooler, S. 1984. "Time pressure, training and decision effectiveness." *Ergonomics* 27:273–284.

Zupanc, C., Burgess-Limerick, R., and Wallis, G. 2007. "Performance as a consequence of alternating control-response compatibility: Evidence from a coal mine shuttle car simulator." *Human Factors* 49:628–636.

Zupanc, C. 2008. *Alternating Steering Control-Response Compatibility: Compatibility, Age, Practice, Strategy and Instruction Effects on Performance Characteristics of Driving a Simulated Underground Coal Mine Shuttle Car*. PhD thesis, University of Queensland. http://espace.library.uq.edu.qu/view/UQ:155665 (accessed April 21, 2010).

Zupanc, C., Burgess-Limerick, R., Watson, M.O., Riek, S., Wallis, G., and Plooy, A. 2009. "A colonoscopy competency framework derived from task analysis." In Shaw, *Proceedings of the 45th Annual Conference Human Factors and Ergonomics Society of Australia. 22–25. Human Factors and Ergonomics Society of Australia, Inc. Canberra.*

Index

F

G

H

For Product Safety Concerns and Information please contact our EU representative GPSR@taylorandfrancis.com Taylor & Francis Verlag GmbH, Kaufingerstraße 24, 80331 München, Germany

T - #0024 - 160425 - C0 - 234/156/13 [15] - CB - 9781439802311 - Gloss Lamination